Contents

Illustrations

Tables

Preface

This study was conducted for the Installation Technology Transfer Program. The project manager was Debbie Lawrence, US Army Engineer Research and Development Center, Construction Engineering Research Laboratory (ERDC-CERL).

The work was performed by Dr. Charles Ryerson, Dr. George Koenig (retired), and Dr. Arnold Song (Terrestrial and Cryospheric Sciences Branch, Janet Hardy, Chief), Kelley MacDonald (Force Projection and Sustainment Branch, Edel Cortez, Chief), and Dr. Donald Albert and Gary Koh (retired) (Signature Physics Branch, Dr. Lindamae Peck, Chief), US Army Engineer Research and Development Center, Cold Regions Research and Engineering Laboratory (ERDC-CRREL); William Stein (Energy Branch, Frank Holcomb Chief, Chief), ERDC Construction Engineering Research Laboratory (ERDC-CERL); Steve Rowley (Operations and Maintenance Division) and Jason Wagner (Environmental Division), Directorate of Public Works, Fort Drum, NY; and Cait Schadock, IMNE-DRM-PWE, Fort Drum, NY.

Funding was provided by the US Army Assistant Chief of Staff for Installation Management (ACSIM) under the Installation Technology Transfer Program (ITTP) under the title "Small Wind Turbine Installation Compatibility Demonstration, Fort Drum, New York." The authors thank Dr. J. Mercer of ALEX Alternative Experts LLC and J. Weale of ERDC-CRREL for insightful and constructive technical reviews of this report.

COL Jeffrey P. Eckstein was the Commander of ERDC, and Dr. Jeffery P. Holland was the Director.

Acronyms

ACSIM	US Army Assistant Chief of Staff for Installation Management
ASOS	Automated Surface Observing System
AWEA	American Wind Energy Association
BLM	Bureau of Land Management
CERL	Construction Engineering Research Laboratory
COTS	Commercial off-the-Shelf
CRREL	Cold Regions Research and Engineering Laboratory
DoD	Department of Defense
DPW	Department of Public Works
DTL	Direct Linear Transformation
EA	Environmental Assessment
EPG	Army Electronic Proving Ground
ERDC	US Army Corps of Engineers Engineer Research and Development Center
ESA	Ecological Society of America
FAA	Federal Aviation Administration
FDED	Fort Drum Environmental Division
HAWT	Horizontal-Axis Wind Turbine
IEC	International Electrotechnical Commission
ITTP	Installation Technology Transfer Program
LRAM	Land Rehabilitation and Maintenance
NAN	Not a Number
NEPA	National Environmental Policy Act
NREL	National Renewable Energy Laboratory
NSI	No Significant Impact

NYSERDA	New York State Energy Research and Development Authority
OSD	Office of the Secretary of Defense
POL	Petroleum, Oil, and Lubricant
PSDvel	Particle Velocity power Spectral Density
RCS	Radar Cross Section
RFI	Radio frequency Interference
RTLA	Range and Training Land
SWCC	Small Wind Certification Council
SWT	Small Wind Turbines
TRL	Technology Readiness Level
UFC	Unified Facilities Criteria
USFWS	US Fish and Wildlife Service
VAWT	Vertical-Axis Wind Turbine
WNS	White-Nose Syndrome
ESPC	Energy Savings Performance Contracts
UESC	Utility Energy Service Contracts
PPA	Power Purchase Agreements

Unit Conversion Factors

Multiply	By	To Obtain
degrees (angle)	0.01745329	radians
feet	0.3048	meters
inches	0.0254	meters
miles (US statute)	1,609.347	meters
miles per hour	0.44704	meters per second
slugs	14.59390	kilograms
square feet	0.09290304	square meters

1 Introduction

1.1 Technical objective

Military bases often ban the construction of large wind turbines, larger than 1 MW in power production, on or near base primarily because of their impact upon the performance of military and weather radars and aviation operations (Rowley 2009; DoD 2006). However, the impact of small turbines, especially those with a capacity of less than 100 kW, on military activities is largely unknown. Even though small turbines are not as efficient as large turbines and they individually provide less power than large turbines, they can collectively contribute to helping the Department of Defense (DoD) and the Army meet national goals of supplying at least 25% of energy on facilities from renewable sources by 2025.

The principal Technical Objective of this Installation Technology Transfer Program (ITTP) project was to assess the compatibility of small wind turbines (≤ 100 kW) with Army installations, using Fort Drum, New York, as a demonstration site. We were to include in the assessment documentation of all known issues with wind turbines of 100 kW and smaller at Army installations (including turbine site locations, endangered species, permitting, etc.). With the assistance of this project, Fort Drum personnel, received permits near the end of FY11 to install one 7.7-kW horizontal-axis wind turbine (HAWT) and one 2.9-kW vertical-axis wind turbine (VAWT) in FY12. The Engineer Research and Development Center (ERDC) and Fort Drum personnel also developed plans to measure radar, acoustic and seismic, turbulence, bird and bat, cold, and icing effects on the turbines and, hence, their effects on garrison operations. The intent of the proposed FY11 through FY13 full project was to indicate how compatible small turbines are with Army installations.

There is currently no Unified Facilities Criteria (UFC) that addresses wind turbines of any size. A product of this project was to be a small wind turbines UFC that would assist energy managers when considering small wind power generation on their facilities. Issues planned for inclusion in the UFC were permitting, environmental, seismic and acoustic, radar, electromagnetic, mission compatibility, location, tower height, visual effects, and economic assessment. The UFC would also include lessons learned from other Army small-turbine installations at Fort Huachuca,

Fort Knox, the Arizona National Guard site, the Navajo training site near Flagstaff, and other sites. We planned to complete the draft UFC by the end of FY13.

This report describes accomplishments made in FY11. Insufficient funding prevented the project from continuing beyond FY11; and therefore, the project did not complete its full objectives.

1.2   Background and problem description

Bases ban large wind turbines on or near DoD facilities when turbines may present a risk to national security, aviation, or endangered species (Sagrillo 2011). In 2004, Fort Drum's Department of Public Works (DPW) Energy Branch proposed installing two 1.5-MW wind turbines on the garrison; and AWS True Wind, under the direction of Pacific Northwest National Labs, made wind speed measurements for one year for that project (Niver 2009). However, though the wind resource was adequate for turbine installation, the Federal Aviation Administration (FAA) denied the installation of the turbines because of radar interference concerns (Rowley 2009).

Though bases ban wind farms on or near DoD facilities when they are anticipated to be an unreasonable national security risk (Seifert 2006), Office of the Secretary of Defense (OSD) reports show that wind turbine impact on DoD missions must be assessed case by case (DoD 2006; NDAA 2011).

For decades, studies have examined the effects of large turbines on radar, humans, and wildlife (Manwell et al. 2002); but little scientifically-substantiated information is known about small wind turbine effects. For example, as with large turbines, small turbines may cause anomalous radar reflections, shadow flicker, bird and bat kills, ice-throw problems, and seismic and acoustic problems for on-base operations or personnel. With little standardization, regulation or oversight claims by manufacturers and marketers that small turbines do not cause these effects are unverified. The many unknowns coupled with no Army-focused guidance make it difficult for garrison managers to judge whether small wind is a reasonable choice for their installation. These unknowns increase risk and potentially cost if managers desire small wind. Additionally, if the facility is risk-adverse, unknowns may decrease options for reaching net-zero energy goals.

Many small wind turbines are available commercially; they are at Technology Readiness Level (TRL) 8 or 9 (Graettinger et al. 2002). However, little is known about their compatibility with Army facilities and Army facility requirements, especially in semi-urban garrison areas where they could interfere with military activities. The commercial availability of small turbines does not in itself imply readiness for use on Army or any DoD facilities. Therefore, from an Army perspective, the TRL of small wind is actually below TRL 8.

Our project proposed to reduce future costs by answering many of the unknowns that slow or stop current attempts to install wind turbines on Army installations. Once this project is completed, the final UFC would enable an expedited process for installation of wind turbines of 100 kW or less. This project has also initiated coordination with a project funded by the US Army Assistant Chief of Staff for Installation Management (ACSIM) at Fort Huachuca to study the electronic interaction of Army systems with commercial scale wind turbines. Coordination between this and the Fort Huachuca project could ultimately allow drafting of a UFC that would address electronic effects of all size turbines on Army systems.

2 Permitting Process

Obtaining permits for wind turbines is complex, time consuming, and can vary by municipality or local authority. The permitting process addresses a wide variety of site-related issues, including turbine location, noise, shadow flicker, proximity to structures, and interruption of radar or communication RF signals. The turbines sites must be in a location that allows maximum energy production based on either data from a wind tower or equally reliable wind model, is compatible with all environmental regulations, shows to have no archeological impact, and is free of any Native American objections. It is also recommended practice to be a "good neighbor" and to reach out to local organizations even if permits are not legally required. Organizations that may require permits or at least good faith notification include local, state, and federal entities. Local organizations may include zoning and planning offices, county boards of commissioners, and city councils. State organizations may include the Public Utilities Commission, state environmental agencies, state historic preservation offices concerned with viewsheds, and industrial development agencies. Typical federal agencies that require permits include the FAA, the Bureau of Land Management (BLM), and the US Fish and Wildlife Service (USFWS) (Juhl 2011). In addition, convention recommends informing the public of installations, especially if the turbines are visible from off-base. Large scale wind turbines are suspected to degrade digital television signals and are known to cause large precipitation and tornado signatures in National Weather Service NEXRAD Doppler radars and television station weather radars (Toth et al. 2011).

Installation of the Fort Drum demonstration turbines required local approval through the release of an Environmental Assessment (Appendix 1), approval from on-base organizations, and permits from federal agencies. Fort Drum approving organizations included the following:

- Environmental Division
- Natural Resources Branch
- National Environmental Policy Act (NEPA) Coordinator
- NEPA Biologist
- Energy Program Manager
- Range Operations

- Ecological Society of America (ESA) Biologist
- Range and Training Land (RTLA) Coordinator
- Hazardous Waste Program Manager
- Wetlands Regulatory Program
- Air Program Manager
- Fish and Wildlife Program Manager
- Petroleum, Oil, and Lubricant (POL) Program Manager
- Cultural Resources Program Manager
- Wetlands Program Manager
- Installation Forester
- Land Rehabilitation and Maintenance (LRAM) Coordinator
- Chief of the Compliance Branch
- Wheeler Sack Airfield Manager
- Plans, Analysis, and Integration Office

The USFWS and the FAA provided permits, and the Department of Defense Energy Siting Clearinghouse provided an approval.

The permitting process is long and can be expensive. At Fort Drum, the permitting process for the demonstration turbines required nearly 10 months. There is also currently a draft US Fish and Wildlife rule that, if passed, would require additional expenditure for studies for any size turbine prior to installation. Facility energy managers must be informed (part of the goal of the UFC) of these site specific laws because they may escalate costs more than available funding levels.

2.1 Fort Drum Environmental Division (FDED) and US Fish and Wildlife Service (USFWS)

The FDED coordinated with the USFWS beginning in January 2011 and conveyed to the ERDC Cold Regions Research and Engineering Laboratory (CRREL) requirements for protecting birds and bats, especially the endangered Indiana bat. CRREL agreed to comply with requirements, including restricted night operation and the use of monopole towers without guy wires (see *Turbine selection* section). The FDED conducted an environmental assessment (EA) to identify and evaluate potential impacts to the natural and human environments from the construction and operation of two types of small wind turbines, one HAWT and one VAWT, to determine the viability of these types of systems for use on Fort Drum and on other Army Installations. EAs address potential impacts to environmental resources such as vegetation, wildlife, threatened and endangered species (particularly relative to this Fort Drum study), soils, climate, air, noise, wetlands and water resources.

In June 2011, the Directorate of Public Works, Environmental Division, and the Natural Resources Branch at Fort Drum prepared The *Environmental Assessment for Conducting a Study of Small Wind Turbines on Fort Drum, New York* (see Appendix 1). The purpose of the EA was to determine the extent of potential environmental impacts from the proposed action and to decide whether or not those impacts were significant, thereby warranting a more detailed study of possible impacts, mitigation, and alternative courses of action. The analysis process involved the review of installation natural-resources-related data collected by Fort Drum and a variety of other governmental agencies and private organizations. The process involved natural resources management, military training and planning, cultural resource management, operations and maintenance, and interviews with personnel from Fort Drum.

The EA concluded with a finding of No Significant Impact (NSI) (Appendix 1). In addition, the EA determined that turbine installation at Fort Drum did not require a Clean Air Act General Conformity Determination or an Environmental Impact Statement. A Public Notice published in local newspapers 3 July through 1 August announced a 30-day comment period with copies of the EA and NSI available for review upon request. The public submitted comments to the FDED, who responded to them accordingly. FDED granted approval in August 2011.

2.2   Federal Aviation Administration (FAA)

Wheeler-Sack Army Airfield management submitted FAA permit applications around 18 May 2011. The FAA conducted an aeronautical study under the provisions of 49 U.S.C., Section 44718 concerning the installation of the HAWT and VAWT. Decisions normally require 30–90 days; and on 14 July 2011, the FAA issued a "determination of no hazard to air navigation" for the 55- and 112-ft turbines (Appendix 2).

The aeronautical study concerning the 55-ft VAWT revealed that the structure did not exceed obstruction standards and would not be a hazard to air navigation. Based on the evaluation, the FAA determined that marking and lighting the 55-ft structure was not necessary for aviation safety. However, if obstructions are marked and lit voluntarily, they should be installed and maintained in accordance with *FAA Advisory Circular*

70/7460-1 K Change 2. The aeronautical study concerning the 112-ft HAWT revealed that the structure did not exceed obstruction standards either and would not be a hazard to air navigation. As a condition to the determination, the study required that the structure be marked or lighted in accordance with *FAA Advisory Circular 70/7460-1 K Change 2, Obstruction Marking and Lighting*—Chapters 4, 5, and 12. It also required that the base complete FAA Form 7460-2, *Notice of Actual Construction or Alteration*, and return it to the Southwest Regional Office—Obstruction Evaluation Group at any time if the project is abandoned or within 5 days after the construction if the structure reaches its greatest height (7460-2, Part II).

2.3   DoD Energy Siting Clearinghouse

Rapid development of renewable technologies has presented many issues among federal, state, and local government compatible use policies and processes. *The National Defense Authorization Act* (NDAA 2011) implemented Section 358: *Study of Effects of New Construction of Obstructions on Military Installations and Operations* to reduce delays of authorization and confusion among all stakeholders. The DoD Energy Siting Clearinghouse, part of the NDAA 2011, Section 358, offers support to installations, regional commands, and services by providing subject matter experts in military testing; training; operations; and radar, sensor, and renewable energy technologies and by providing a science-based evaluation of project proposals that are likely to have significant impacts on military missions. A focal point of the Clearinghouse is to help services develop analysis tools and share best practices. Fort Drum's Plans, Analysis, and Integration Office submitted the CRREL ITTP project plans to the DoD Energy Siting Clearinghouse for review and approval in April 2011. Fort Drum indicated no local concerns in the submission; therefore, the Clearinghouse found no reason to review the project, and it was approved (Appendix 3). Only projects with significant impacts or that need multi-service coordination receive full Clearinghouse attention.

3 Turbine and Demonstration Site Selection

In 2004, the FAA denied Fort Drum's installation of two 1.5-MW wind turbines because of potential radar interference. Fort Drum has since considered installing many small turbines of similar net capacity because they may not cause significant radar interference. We selected Fort Drum for demonstration because of this interest in small turbines.

The Fort Drum facility Commander has a record of denying requests to install turbines on base because of apparent conflicts of interest when he denies installation of turbines off-base that may interfere with radar and flight operations. However, the facility Commander has supported this project because of its potential to benefit the entire Army in helping meet renewable energy goals.

3.1 Turbine selection

One must consider several factors before selecting a turbine for installation at any site, including type of turbine (VAWT or HAWT), power output, aesthetics, noise, lot size, and foundation requirements. We designed this study to demonstrate the impacts of small wind turbines (SWT) on Army facility operations; it was not intended to demonstrate the performance of a wide range of turbine models. However, selection of the two demonstration turbines was constrained by the preferences of Fort Drum approving organizations and the ERDC research team and by requirements of the USFWS.

The FDED conducted a teleconference with ERDC on 4 February 2011 indicating that the USFWS had granted preliminary approval for the project to proceed if it met the following criteria for turbine selection (FDED 2011, Appendix 4):

1. The turbines could not operate when bats, especially the endangered Indiana bat that has on-base habitat, are most active because of the danger of collision with the moving turbine blades. Therefore, turbine blades were not allowed to turn from 1800 to 0800 hr daily between April and October for the first year of the project. This criteria required a mechanical brake to completely stop the rotation of the blades

This constraint to restrict operation of the turbine for 14 hr would result in an inoperable machine for more than half the allowable time in a 24-hr period, and the restriction was for 60% of the year. This is an important consideration when evaluating the economic payback of the turbines, especially during the first year of the project.

2. To reduce roosting or nesting locations for birds and bats near the turning turbine blades, the turbines could be mounted only on monopole towers without guy wires. The guy wires and spinning blades could result in fatalities. In addition, no other appurtenances would be allowed, such as arms to hold anemometers for turbulence measurement, on the turbine towers or on any other nearby tower, which may also provide bird or bat roosting places.

   Monopole towers are generally more expensive than lattice towers. The selection of a monopole tower would add an additional cost to the overall project budget; and therefore, this would impact the economic payback period.

3. Steady burning lights can attract birds and bats; therefore, the FDED requested that no steady burning lights be mounted on the turbines. If lighting was required, FDED requested the use of white strobe lights with the longest possible off phase (at least 3 s off). Towers and turbines less than 200 ft above ground level do not require clearance lights by the FAA, but Army aviation training operations near ground level on Fort Drum required that the turbines be visible at night; to meet this criteria, the project planned for a slow flashing white strobe. The placement and type of lighting did not affect the selection of the turbines. We verified with the distributor and manufacturer of each turbine that FAA and security lights are capable of being mounted on the turbines.

4. All ground-level security lights must be down shielded to reduce attraction of birds and bats.

In addition to the above FDED and USFWS requirements, Fort Drum and ERDC preferred that the turbine monopole towers be at least 90 to 100 ft tall to allow better wind exposure and that the towers tilt up or down using a gin-pole. Turbines mounted on tilting towers typically do not need high-cost cranes for their installation, maintenance, or dismantling.

Fort Drum was most interested in installing vertical-axis turbines. Manufacturers claim that VAWTs are more efficient at lower heights, more resistant to turbulence effects, quieter because they rotate at slower speeds, and less dangerous to wildlife because of their slower rotation speed compared to the more common HAWTs. However, HAWTs are more common, are supposedly more efficient, and are generally smaller versions of the megawatt and larger turbines used in wind farms. Therefore, the project chose a VAWT and a HAWT for demonstration and comparison at Fort Drum.

ERDC began a market survey of commercial off-the-shelf (COTS) turbines within the 3–10 kW range in mid-November 2010. We conducted a general internet search to begin the survey of the 400+ models of small HAWTs (www.allsmallwindturbines.com) and the fewer than 100 VAWTs. We collected specifications and documentation from manufacturers and distributors of turbines (see Appendix 5 for list of turbines from the market survey) that fell within the acceptable power output range. Specifications assessed included the following:

1. Turbine model number or name
2. Power rating at 11 m/s (about 36 ft/s)
3. Tower type (lattice, monopole, tilting monopole, etc.)
4. Tower material
5. Rotor diameter
6. Blade speed
7. Usable wind speed range (cut-in, cut-out, and survival speeds)
8. Bird and bat kill documentation
9. Documented noise and vibration measurements
10. Radar impact measurements
11. Blade material
12. Deicing and anti-icing capability
13. Certification by independent organization (Small Wind Certification Council [SWCC], National Renewable Energy Laboratory [NREL], etc.)
14. Positive braking system
15. Cost, delivery time, ordering lead time
16. Expected life of turbine
17. Dealer and manufacturer locations
18. Customer contacts and location of installed turbines
19. Warranty

From the initial market survey, we compiled a short list of acceptable tur-bines meeting requirements set by the FDED and FDWSA, and ERDC fi-nalized the list in mid-February 2011. The list consisted of the four HAWTs and three VAWTs listed in Table 1:

Table 1. Acceptable Turbines Selected for Consideration.

HAWT	VAWT
Aerostar 6 m	CleanField V3.5
Bergey EXCEL-S	Urban Green Energy 4K
Proven 11	UrWind O_2
WindTamer 4.5/8.0 GT	

On 12 April 2011, Fort Drum's Master Planner, the Public Works Director, and the FDED indicated that it desired a down-select to one HAWT and one VAWT. ERDC executed the down-select by the week of 25 April 2011 and paid close attention to the braking system and the tower options and capabilities. ERDC selected the Aerostar 6 m (HAWT) and the CleanField V3.5 (VAWT) (Fig. 1). We then made a formal request to Fort Drum that it submit permit applications to the USFWS and to the FAA for the selected turbines at the approved site.

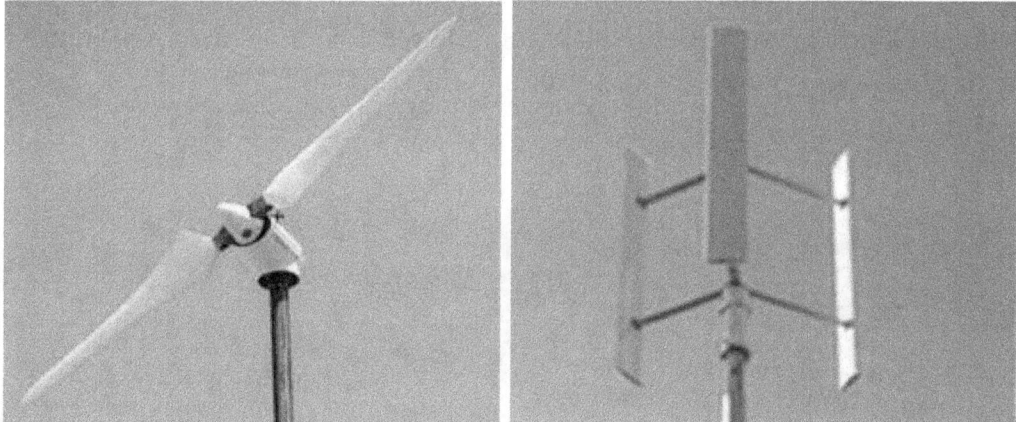

Figure 1. Aerostar 6-m horizontal-axis turbine (left) and Cleanfield V3.5 vertical-axis turbine (right).

3.2 Site selection

Fort Drum is located in northern New York State, immediately east of the eastern end of Lake Ontario and immediately west of Adirondack Park. The Fort's center is approximately 44°7' N, 75°35' W and is approximately 10 mi (16 km) wide and 20 mi (32 km) long (Niver 2009) (Fig. 2).

Photo Removed Due to Copyright Restrictions

Figure 2. Map compiled and modified from Figures 2 and 3 of Niver (2009). Red circle
indicates the approximate area of Figure 3 below with all of the considered turbine sites.

We initially considered three sites at Fort Drum for the two demonstration
turbines and inspected these sites on 15 November 2010. All of the sites
were within the main cantonment area or on the western edge of the
Wheeler-Sack Army Airfield—all within the red oval in Figure 2 and num-
bered 1, 2, and 3 within the red circles in Figure 3.

Photo Removed Due to Copyright Restrictions

Figure 3. Locations of Candidate Sites, former wind tower, and endangered Indiana Bat
habitat. Map compiled and modified from Niver (2009) Figures 4 and 5.

Candidate Site 1 is on the northwest side of Childers Indoor Range (P-11160) on a small ridge (44°03.793' N, 75°47.644' W) (see red circle with "1" in Fig. 4). The area has a dense tree cover that is approximately 40 to 60 ft high. Though a meteorological tower had been placed southwest of the site about 6 years earlier to assess the wind available for 1.5-MW turbines (see red circle with "T" in Fig. 4), the site has subsequently been found to be on the edge of the endangered Indiana bat habitat on the Fort (Niver 2009) (Fig. 4). In addition to the habitat restriction, the site was judged a potentially poor wind site because of dense tree cover (Fig. 4). However, the site may be acceptable for turbines on higher towers.

Figure 4. Views of Candidate Site 1. Upper panel view from left to right is west through north to east. Lower panel left to right is east through south to west.

Candidate Site 2 is on Division Hill in a parade ground area (44°02.127' N, 75°44.695' W) (see red circle with "2" in Fig. 3). Division Hill, the former site of structures and currently carrying powerlines, is a prominence standing at least 20 ft higher than the surrounding open, relatively flat parade ground. The entire area is nearly treeless (Fig. 5). Flagging (tree branches on the windward side of the tree are deformed or killed, giving the tree a flag-like appearance) of nearby needle-leaf trees suggests persistent, strong winds (Fig. 6). However, the site was ultimately considered unusable because it is too near the southern approach and departure end of the Wheeler-Sack Army Airfield.

Figure 5. Views of Candidate Site 2. Upper panel view from left to right is north through east to south. Lower panel left to right is south through west to north.

Figure 6. Tall flagged trees in foreground near Candidate Site 2 suggest strong, persistent winds from left to right (winds from the west).

Candidate Site 3 is located at the Outdoor Wash Facility (P-21510) (see red circle with "3" in Fig. 3). The Outdoor Wash Facility is for washing vehicles and is used relatively infrequently. It is located immediately west of the Wheeler-Sack Army Airfield, and a 200-ft water tank shadows it from the airfield radar (Fig. 7). There is also a 200-ft cell tower located between the site and the airfield.

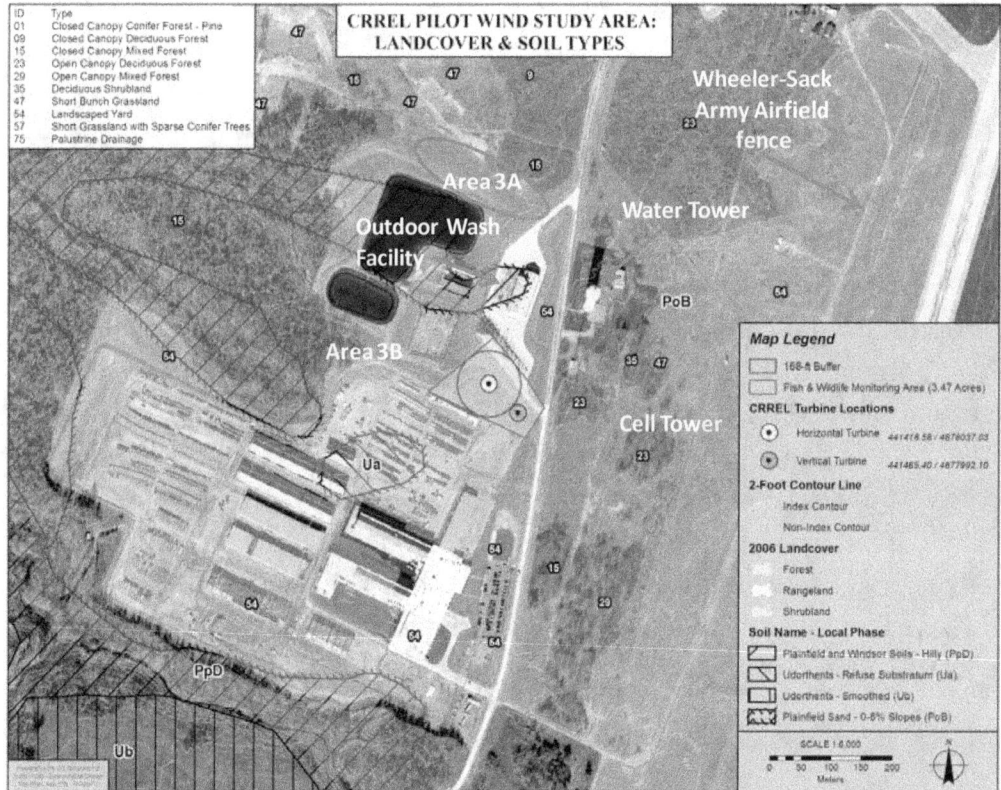

Figure 7. Candidate Site 3, with Areas 3A and 3B and locations of water tower, cell tower, wash facility, and airfield.

We further considered for the turbines two locations within Candidate Site 3. The first, at area 3A (Fig. 7), is nearly north of the wash facility and sits lower in elevation than the wash facility, which is on an approximately 20-ft ridge. However, space on the site was constrained by potential new construction to the north and the reservoir to the south.

Area 3B (Fig. 7) was the location we chose for the turbines; and it was approved by Fort Drum, the Wheeler-Sack Army Airfield, the FAA, and the USFWS. Though somewhat physically constrained by limited space, the turbines fit on the site with space available around each to meet the Fort Drum Environmental Division bird and bat carcass retrieval area requirement to have a circular area of clear land surrounding the base of each turbine with a radius equal to 1.5 tower heights. A powerline is located between the Outdoor Wash Facility and the site (Fig. 8). Soil conditions are sandy and appropriate for turbine foundations. Coordinates for the turbines are HAWT 44°03.187' N, 75°43.889' W, and VAWT 44°03.165' N, 75°43.849' W.

Figure 8. Area 3B, selected for the turbines, viewing to the northwest.

The turbine site was for demonstration purposes only and was not selected for optimal wind conditions. The New York State Energy Research and Development Authority (NYSERDA) rated the winds at the site "average"—about 11 to 12 mph at the 100-ft hub height of our HAWT. NREL places our site in an annual Class 3 (Moderate) Power Density ranging from a summer Class 1 (Low) to a winter Class 4 (Good). Our analysis of airfield wind data with a 0.29 wind shear exponent representing "trees, hedges, and a few buildings" (Gipe 2004) shows that the 100-ft Aerostar 6-m HAWT should provide an annual energy output of 12,323 kWh. The 50-ft VAWT should provide an annual energy output of 2225 kWh.

4 Radar and EMI Signature Background and Methodology

One significant goal of this project was to assess the effects of small turbines on Doppler radars used on many Army facilities. The effects of large wind turbines on radar systems are well-documented (Toth et al. 2011; NOAA 2011; Seifert 2006; DoD 2006). Wind turbines create static clutter from the radar cross section (RCS), dynamic clutter from the Doppler signal reflecting from rotating parts, and shadowing or blocking of areas behind the turbines. We planned to measure the effect of the small turbines on radar signatures during field tests at Fort Drum. In preparation for the field work, we conducted laboratory experiments with scale models and collaborated with the Army Electronic Proving Ground (EPG) at Fort Huachuca. We planned to compare the RCS and Doppler signals from the small HAWT and VAWT with an EPG 1-MW Nordic Windpower Model N1000 2-bladed HAWT, similar in design to the Aerostar 6-m turbine (Fig. 9). We began discussions with the EPG at Fort Huachuca and visited the site in 2011, and we conducted laboratory experiments with model turbines at CRREL.

Figure 9. Nordic Windpower Model N1000 1-MW 2-bladed HAWT at Fort Huachuca.

We conducted controlled laboratory scale model tests using a vector network analyzer operating up to 40 GHz. These tests incorporated a variety of incident and azimuthal angles to the scale model turbines and were at varying blade rotation speeds. We intended to duplicate these measurements on the Fort Drum turbines and to compare Doppler signals to blade rotational speed and the HAWTs azimuthal position fixed in relation to radar position. Determining the characteristic Doppler signatures through signal processing may allow us to reduce wind turbine signal clutter.

A Doppler radar operating at 1.5-GHz (20-cm wavelength) center frequency was constructed to collect the radar backscatter measurements from the full-scale wind turbines at Fort Drum. We selected this radar frequency because many Doppler radars fielded by the FAA, DoD and NOAA operate at or near 1.5 GHz (Fig. 10). The microwave source for the radar was a HP8350B sweep oscillator with a HP83522A plug-in module. The transmit and receive antenna for this system was a 15-dB standard gain horn antenna whose 3-dB beam width in the E- and H-planes was 31° and 20°, respectively. A circulator and a low noise amplifier collected and amplified the Doppler signals produced by the motion of the rotating blades. We initially tested the radar in the laboratory using a room fan with 16-in. plastic blades. The radar easily detected the rotating blades even though the plastic blades were expected to have a low radar cross section. We did not complete additional work with this radar in FY11 because the turbine installation at Fort Drum was delayed until FY12.

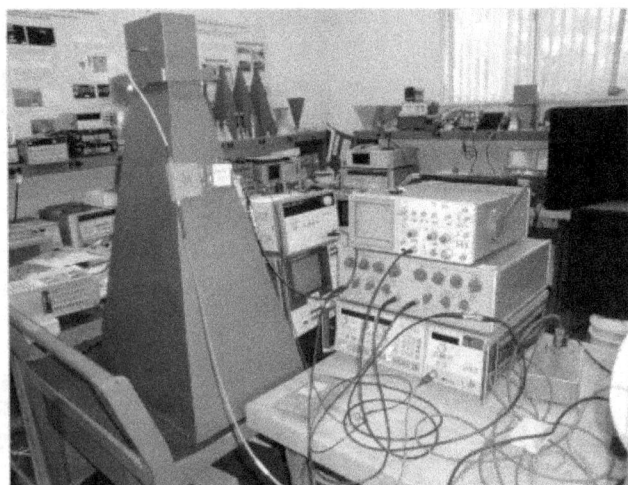

Figure 10. 1.5-GHz Doppler radar built to evaluate turbine radar reflections at Fort Drum.

The scaled laboratory studies required a Doppler radar operating at a much higher frequency (shorter wavelength). Therefore, we constructed a

Doppler radar operating at 95 GHz (0.3-cm wavelength) (Fig. 11) and used a HP8350B sweep oscillator with a HP83590A plug-in module to generate a continuous microwave signal at 1.58 GHz. This signal then passed through a Millitech frequency multiplier (6×) to generate the 95-GHz radar signal. A 40-dB gain-horn lens antenna with a beamwidth of 2° transmitted and received the radar signal.

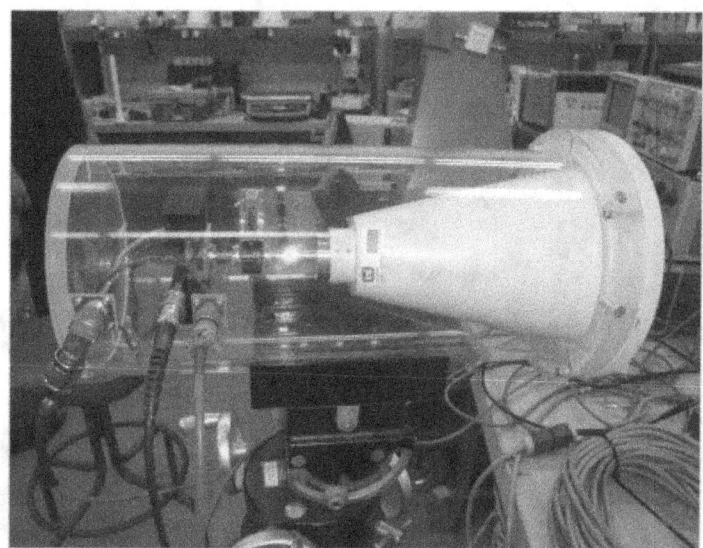

Figure 11. 95-GHz Doppler radar built to evaluate scale model turbine radar reflections at CRREL.

Lionel, a manufacturer of model trains and accessories, provided the scaled wind turbines used for this investigation. These wind turbines were accurately modeled and had three spinning blades, a nacelle, and a tower. The turbine was approximately 40 cm tall and the blades measure 21 cm in length. At 1:66 scaling, a model turbine represented an actual wind turbine footprint of approximately 28 m in diameter. The turbine rotated approximately 0.3 rotations each second. This represented an angular velocity of approximately 38 cm/s at the tip of the blades. For a Doppler radar operating at 95 GHz, the maximum Doppler frequency expected was to be no more than 250 Hz. Therefore, we could easily record these raw signals with our National Instrument data acquisition card.

The material properties and the shape of these model turbines were not the same as those selected for the full-scale wind turbines at Fort Drum. Therefore, the results obtained using these scale models cannot be used to predict the radar reflection from real turbines. However, the measurements using these model wind turbines provided valuable insight that one can use to develop field testing strategies to better understand radar-wind turbine interaction. If a realistic scale model of a turbine (same material and shape) were constructed, laboratory testing could be used to obtain the necessary information about the radar-wind turbine interaction. These measurements would reveal the type of radar interference that one could expect. This knowledge was required to develop potential mitigation strategies.

Figure 12 illustrates the 95-GHz radar pointed in the direction of several model turbines. Radar absorbing foam lined the background to reduce multiple radar reflections. The initial testing investigated reflection from a single turbine.

Figure 12. 95-GHz Doppler radar aimed at four rotating model turbines.

We measured radar returns from a rotating turbine at two different yaw angles (the relative orientation of the wind turbine and radar). A yaw angle of 0° means that the turbine is facing the radar. A 90° yaw means that the turbine is facing in the direction perpendicular to the radar line of site. Figure 13 illustrates radar returns from the three blades after two full rotations. The top and bottom plots represent measurements from yaw angles of 50° and 70°, respectively. Reflections from each blade are shown twice and are separated in time by approximately 3 s.

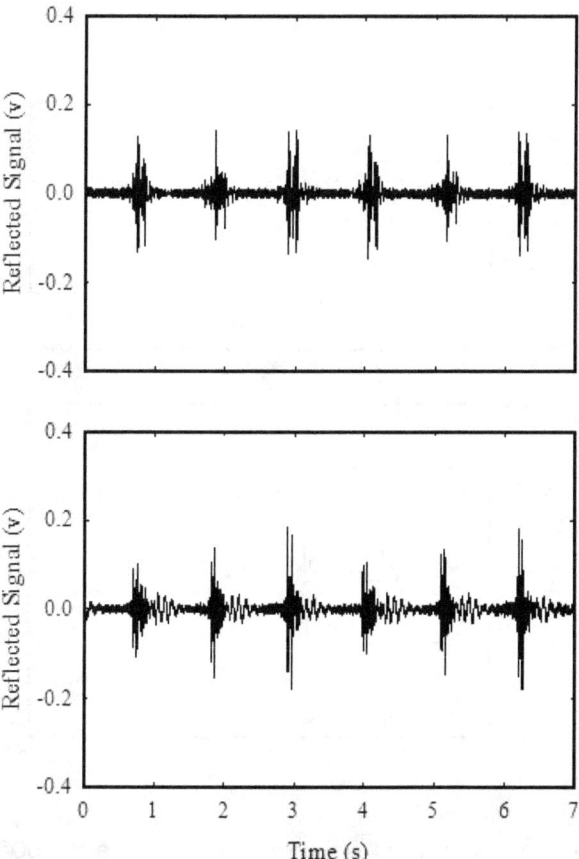

Figure 13. Doppler radar returns from two rotations of three blades. Upper image represents a yaw angle of 50°, and the lower image is a yaw angle of 70°.

The radar backscatter from each blade is very repeatable. However, there are differences among the three blades. This suggests that the three blades are not constructed uniformly. The radar reflections from a rotating blade are also not a continuous event; each blade produces a radar "flash" as it spins. This flash occurs when the blade and radar antenna are aligned such that the blade returns a specular reflection.

Figure 14 shows a detailed view of the radar returns from a single blade. The radar return from the top figure (50° yaw angle) has a lower Doppler frequency than the bottom plot (70° yaw angle). This is expected because the radar measures the velocity component in the direction of the radar. Figures 13 and 14 suggest that two distinct reflections are present from a blade (regardless of yaw angle). The mechanism responsible for these returns is unresolved. Future laboratory studies using scaled turbine blades that are constructed using the same material and pitch as the larger turbines can help to gain important information about radar wind-turbine interaction without resorting to expensive field testing.

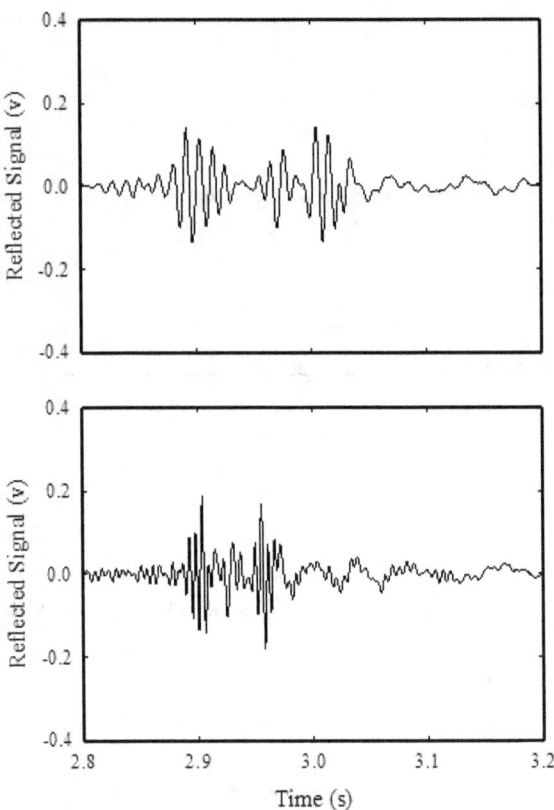

Figure 14. Doppler frequency velocity component from a blade at a 50° yaw angle (upper image) and at a 70° yaw angle (lower image).

5   Seismic Acoustic Signature Background and Methodology

Acoustic noise produced by large wind turbines has been studied for years, but there is still controversy over human annoyance parameters—the traditional noise exposure criteria (a noise level integrated over all hearing frequencies, i.e., a single number) developed for traffic and aircraft noise underestimates the impact of intermittent wind turbine noise. In addition, there have been few studies of vibrations produced by wind turbines because the vibration levels a few hundred meters from the wind towers are always below human perception levels. Further studies and documentation is necessary for small turbines because they are more likely to be used in large numbers near people or mounted on buildings where vibration may cause annoyance. We have learned that AWEA (American Wind Energy Association) small wind certifications are now requiring acoustic sound testing (determining sound pressure levels at integer wind speeds) (AWEA 2009a; Colby et al. 2009; IEC 2006) for each turbine.

Large turbines are generally louder than small turbines because of their size. However, per megawatt, small turbines are often less well-acoustically insulated, are closer to the ground, and are closer to people. Small wind turbine noise has been assessed using methods developed for all sizes of wind turbines by the International Electrotechnical Commission (IEC 2006). However, there are currently no small turbine noise standards in the United States, and IEC standards for small turbines are argued in the United States to be inadequate because they do not capture noise fluctuations.

5.1   Fort Drum observation plan

We anticipate that small turbine sound and vibration (compared to large turbines) will be lower in magnitude but higher in frequency, where human hearing is more sensitive. Therefore, we proposed to conduct experimental measurements of both acoustic and seismic noise levels produced by the small wind turbines at Fort Drum as a function of frequency and wind speed. We would use IEC standards as guidelines but would expand them to record the noise produced by the turbines at selected distances and seasons over an extended frequency range. Seismic and acoustic measurements on a Northwind 100-kW (NW100) turbine at NREL in 2006 (Albert and Perron 2006) using our methods (explained below) revealed three strong frequencies emitted by the tower, hub rotation, and the three blades passing the tower.

Using integrating sound level meters, we planned to measure noise levels at Fort Drum in one-third octave frequency bands from 6 to 20 kHz for acoustic noise and from 6 to 1000 Hz for vibration. This is a standard technique in acoustics (but not for vibration) and is often used to assess traffic or industrial noise. However, human annoyance seems to be related to the fluctuations in the noise, and these variations are not well captured by standard measurements. Therefore, we had planned to make continuous time series measurements to examine fluctuations and impulsive noises that seem to be responsible for noise complaints. For these measurements, we would have installed an array of microphones and vibration sensors at a few distances from the wind turbines and digitally record the noise with GPS-accurate timing. These measurements should have fully captured the variability and fluctuations in the noise produced by the turbines, and they could have been correlated with the turbine operating status (chiefly rpm and electrical output) and with wind speed. In many cases, the noise of small turbines changes with power output because of changes in load on the blades.

5.2 Previous wind turbine measurements

We have conducted similar measurements on large wind turbines in the past with different equipment to account for the lower frequency output of the larger devices. These earlier measurements served as the fundamental proof-of-concept methodology for the Fort Drum field work. The previous work showed that the generator emits three frequencies (Fig. 15). The strongest signal was at a frequency of about 1.3 Hz with a particle velocity power spectral density (PSDvel) of 1×10^{-10} (m/s)^2/Hz at a distance of 3 m. Because its frequency did not change with blade rotation rate, this signal is likely associated with the fundamental frequency of the tower supporting the wind generator. The next frequency varied with the hub rotation speed and was usually around 0.8–1.0 Hz with levels reaching 2×10^{-11} (m/s)^2/Hz. The frequency was at the blade passing frequency, usually around 2.5 Hz, and had a spectral density also around 2×10^{-11} (m/s)^2/Hz. These spectral lines produce an RMS vibration level of about 5×10^{-6} m/s. The bottom panel of Figure 15 shows the spectral levels of these lines.

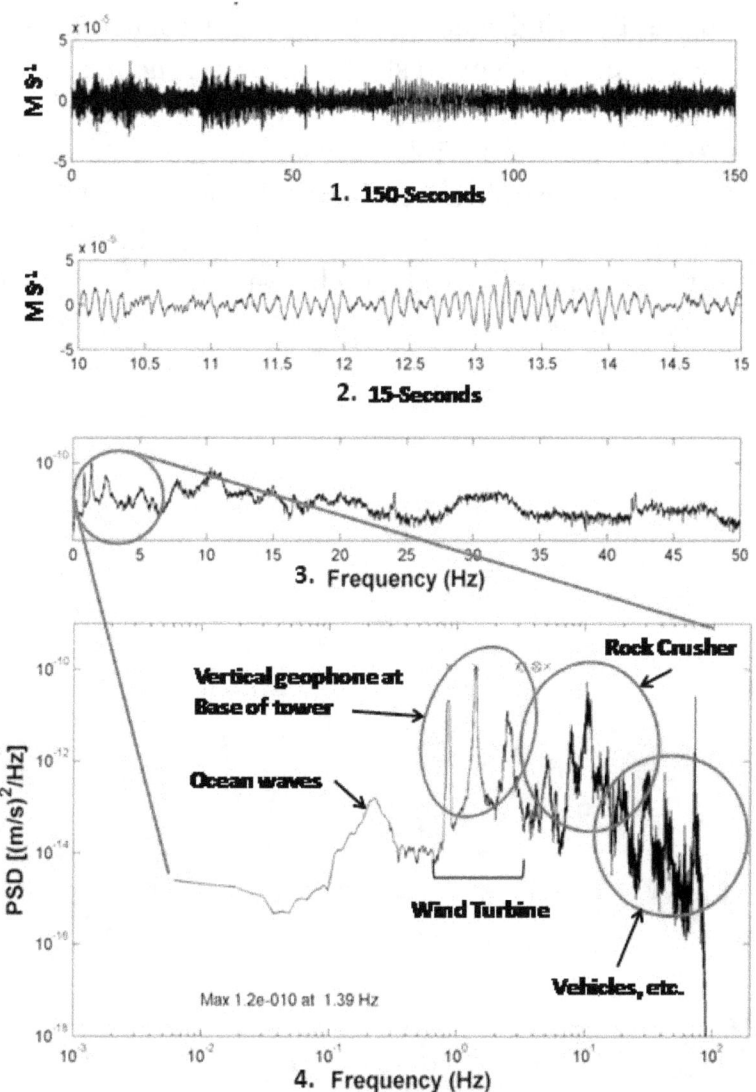

Figure 15. Vibration measurements made at NREL 25 April 2005 at about noon of Northwind 100-kW turbine. Three-blade turbine rpm is 51.7, wind speed is 2.0 m s⁻¹, and RMS vibration is 4.8 × 10⁻⁶. Top to bottom panels show (1) complete 2.5-min time-series, (2) a 5-s time series detail, (3) the low frequency portion of the spectrum on a linear vertical scale, and (4) the broad band logarithmic power spectrum. The three strong peaks at 1, 1.5, and 2 Hz are produced by the wind turbine.

The low frequency peak at 0.23 Hz in the lower panel of Figure 15 is the microseism peak and is produced by waves along the west coast of the United States and water waves at sea. This peak was visible in all of the spectra we measured, but we could not detect it in the higher frequency measurements that would be used for the smaller Fort Drum wind turbines. We also recorded strong vibrations (at 3.1, 4.1, and 5 Hz) from an industrial plant located about 700 m away from the turbine and from automobile traffic nearby.

We also noted that the wind generator occasionally rotated to face the wind. This yawing action produced strong vibration and a different signature with two strong spectral peaks (at 1.1 and 1.4 Hz) at a level of nearly 1×10^{-10} (m/s)^2/Hz. If the wind is relatively steady in direction, these signatures will not be encountered frequently nor should the VAWT produce them. Finally, because of the very strong signals from the off-site industrial operations, the anticipated correlation between the vibrations and the wind speed did not occur.

6 Turbulence Effects Background and Methodology

Our objective is to understand the performance of small turbines in response to changing wind flow patterns and turbulence in a semi-complex environment. Wind turbines operate most efficiently in winds that are laminar in flow. Turbulence caused by terrain, obstacles, and weather can reduce power output, stress the wind turbine and tower, and increase maintenance and reduce system longevity. Flow over idealized flat uniform terrain is horizontally homogeneous (partial derivatives of mean parameters in the horizontal plane can be ignored) and exhibit stationarity (time derivatives vanish). In the real world, idealized flat terrain surfaces do not exist. The closest we can come to these idealized conditions occur in flat regions with uniform low vegetation—for example, the plains in Nebraska, Kansas, and Minnesota.

At the proposed wind turbine tower sites at Fort Drum, variations in the terrain characteristics and nearby canopy and buildings will result in non-laminar flow (Fig. 16). In addition, varying terrain can cause differential solar heating resulting in turbulence.

Photo Removed Due to Copyright Restrictions

Figure 16. View of the area surrounding the proposed location of the two wind turbine towers (image generated via Google Maps).

The terrain and obstacles near the proposed wind turbine site consist of a forest canopy located to the southwest, west, and north of the towers; buildings to the south and southwest; and an airfield located to the east. A close-in view (Fig. 17) reveals several structures near the proposed wind turbine locations, including a water tower approximately 700 ft to the northwest. Canopy, buildings, and changing topography disrupt the wind flow and can result in turbulent conditions that reduce the efficiency of the wind turbines.

Photo Removed Due to Copyright Restrictions

Figure 17. Close-in view of the potential location of the wind turbines (image generated via Google Maps).

6.1 Turbulence monitoring

Had this project continued, we would have characterized the environmental conditions to understand how the performance of the VAWT and HAWT change with changing conditions. The measurement of the environmental conditions would entail erecting a small 6- or 9-ft meteorological tower. This tower would contain instruments to measure the air temperature, relative humidity, pressure, visibility, and wind speed and direction. We would also obtain upwelling and down-welling infrared and solar flux measurements. To characterize precipitation events, we would operate a rain gauge and an ice detection system. We would supplement these measurements by obtaining the Wheeler-Sack Army Airfield measurements, if available.

Our main interest would be to characterize the winds and turbulence near the tower. To achieve this objective, we would operate at least four cup anemometer systems at shelter height (about 2 m above ground level) at select locations. We would also operate sonic anemometers capable of measuring the three components of the eddy velocities associated with turbulent flow. These locations would be based on a survey of the topography, the location of the canopy and buildings relative to the tower locations, and the prevailing wind direction. Analysis of the Wheeler-Sack wind data (Fig. 18) indicated the prevailing wind directions for most months are from the SW, SE, and NE. The cup anemometers would operate at the highest possible sampling rate to infer turbulence. To obtain information on the vertical structure of the wind field and turbulence, we would operate sonic anemometers at three elevations on a nearby cell tower, provided we were granted permission to use the tower (Fig. 19). The sonic anemometers could affix to 6-ft extensions mounted perpendicular to the axis of the tower and not require guy wires. We would locate the sonic anemometers on the prevailing upwind side of the tower. The vertical spacing of the sonic anemometers is usually logarithmic. We would locate four additional cup anemometers, operating at the highest possible sampling rate, on the cell tower at the same elevation as the sonic anemometers.

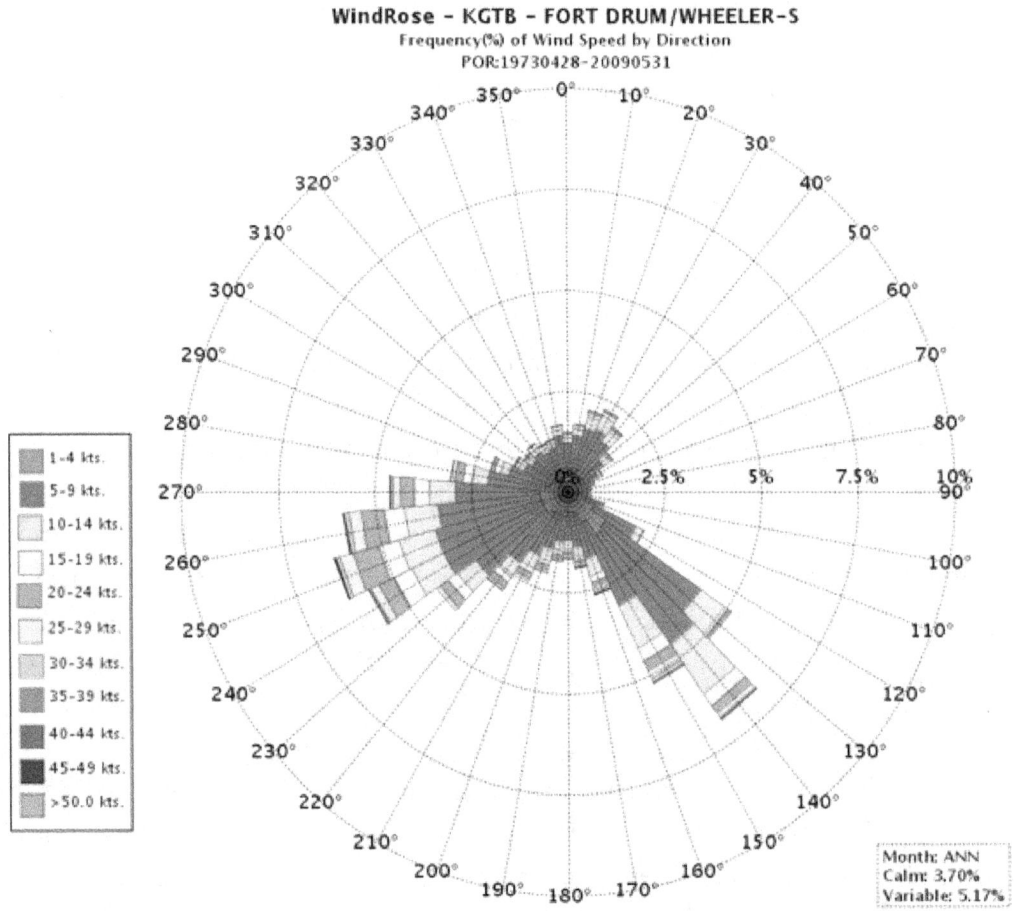

Figure 18. Annual wind rose calculated from monthly averages of observations from 1973 through 2009 at Fort Drum/Wheeler-Sack Army Airfield.

Figure 19. Location of the cell tower, the Outdoor Wash Facility, and the turbine site 3B. The VAWT would have been located approximately below the left arrowhead under the Area 3B label, and the HAWT would have been located approximately beneath the right arrowhead. View is from north of the Outdoor Wash Facility near Area 3A towards Area 3B, from NNE to SSW.

6.2 Turbulence monitoring methodology

Cup anemometers can be potentially cheap alternatives to sonic anemometers. While sonic anemometers range in price from $1500 to $2500, cup anemometers range in price from $250 to $400. However, the high sampling rate, approximately 10 to 30 Hz, required to measure eddy velocities is not available with conventional pulse cup anemometers. The NRG 40C cup anemometer uses the Hall Effect that results in a continuous voltage output. The frequency of the time varying voltage is a linear function of the wind speed. Preliminary measurements indicate it may be possible to sample the NRG 40C at a rate that will provide eddy velocity information.

As a proof of concept experiment prior to conducting the Fort Drum field work, we placed a sonic anemometer and NRG 40C anemometer in an open area and obtained measurements (Fig. 20). The sampling rate for the NRG 40C was 10 s (0.1 Hz), while the sampling rate for the sonic was 0.1 s (10 Hz). It is evident from Figure 20 that the 10-s sampling rate is insufficient. NRG data collected at a different time with a sampling rate of 2 s (0.5 Hz) captures more of the higher frequency variations but still does not capture some of the higher frequency variations in the eddy velocities. The NRG operated at sampling rates of 0.1 and 0.5 Hz is not able to resolve some of the small-scale variations in the eddy velocities as seen in the trace for the sonic anemometer. Because of the different sampling rates, the average, maximum, minimum, and standard deviation of the sonic and NRG eddy velocities differ as indicated in Table 2. Visual inspection of the two graphs reveals the NRG plot is smoother and does not capture high frequency variations in the eddy velocities.

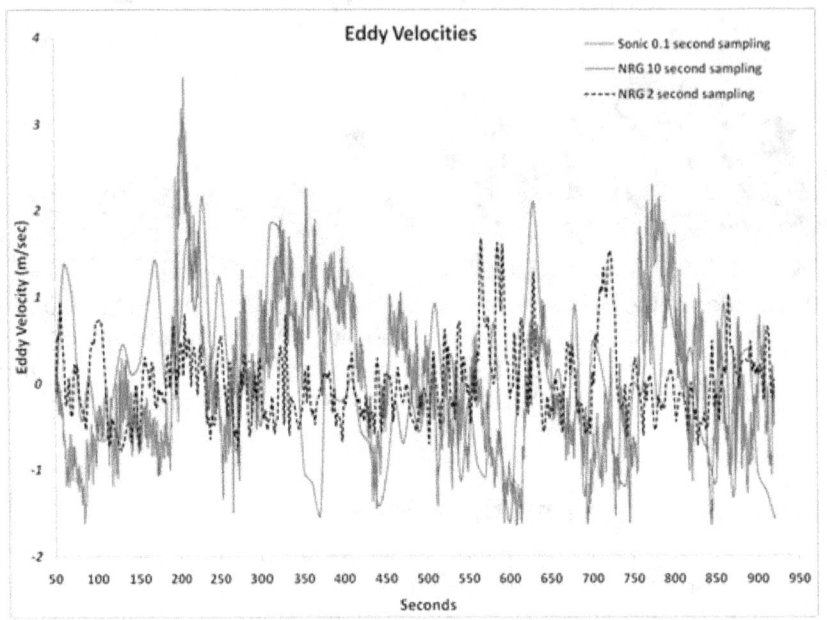

Figure 20. Comparison of eddy velocities for sonic and cup anemometers with different sampling rates. The Sonic and NRG (NRG Systems is a manufacturer of cup anemometers) (10-s sampling) eddy velocities are for the same time.

Even though the NRG data collected with a 2-s sampling rate is for a different 15-min period, it is still evident that even at this higher sampling rate the NRG is not capable of capturing high frequency variations in the eddy velocity.

Tables 2 and 3 present the statistical variables for the sonic and NRG data. For the sonic and NRG data collected for the same time, we computed the gustiness defined as

$$g = \frac{\overline{|u'|}}{U}$$

where U is the average velocity over the period, and u' is the eddy velocities defined as

$$u' = u - U$$

and u is the sampled velocity at a given time. The turbulence intensity is defined as

$$I = \frac{\sqrt{\overline{u'^2}}}{U}$$

Table 2. Velocity (m/s) statistics for the sonic and NRG anemometers for a 15-min data period.

Statistic	Sonic (0.1 sec sampling)	NRG (10 sec sampling)	NRG (2 sec sampling)
Average	1.67	3.05	2.01
minimum	0.02	1.46	1.23
maximum	5.12	5.21	3.68
Standard deviation	0.86	0.88	0.44

Table 3. Eddy velocity (m/s) statistics for the sonic and NRG anemometers for a 15-min data period.

Statistic	Sonic (0.1 sec sampling)	NRG (10 sec sampling)	NRG (2 sec sampling)
minimum	-1.64	-1.59	-0.78
maximum	3.54	2.56	1.67
Standard deviation	0.86	0.88	0.45
gustiness	0.42	0.22	NA
Turbulent Intensity	1.13	0.23	NA

To use the cup anemometer to capture the eddy velocities associated with turbulence, it is necessary to sample the output from the anemometer at the highest possible rate. As indicated, the NRG 40C anemometer outputs a frequency rather than a pulse as is the case with many anemometers. The output frequency is linearly proportional to the wind velocity and can range from 0 to 125 Hz. The way the system (anemometer and the Campbell data logger) is configured, the highest possible sampling rate is approximately 5 Hz. The Campbell data logger timeout value for this sampling rate is 190 ms. This means there must be at least one cycle during the 190 ms to compute a wind speed. A single cycle in 190 ms corresponds to a frequency of 5.26 Hz. Using the equation vel = 0.765 × freq + 0.36 given by NRG for computing the wind speed from the frequency information, the corresponding velocity is 4.38 m/s (9.8 mph). The HAWT cut in wind speed is 3.57 m/s while the cut in wind speed for the VAWT is approximately 5.0 m/s. Thus, operating the NRG anemometers at 5 Hz does not support the operating range for the HAWT for wind speeds below 4.38 m/s, resulting in a wind speed recorded as NAN (Not A Number).

It is possible to reconfigure the system to use a lower sampling rate of approximately 2 Hz that corresponds to a timeout value of 450 ms and a lower wind speed limit of 2.01 m/s. However, the lower sampling rates and longer timeout value limit the capability to measure the higher frequency variations in the wind speeds as illustrated by the theoretical example presented graphically in Figure 21. In this example, the frequency of cup rotation for the first 225 ms is 41 Hz, corresponding to a wind speed of 32 m/s. For the next 225 ms, the frequency is 13 Hz, corresponding to a wind speed of 10 m/s. However, the timeout window is 450 ms; so the frequency over this window is 26 Hz, corresponding to a wind speed of 20 m/s. While increasing the timeout value enhances the ability to record lower wind speeds, it does not resolve changes in wind speeds over time intervals less than the length of the timeout.

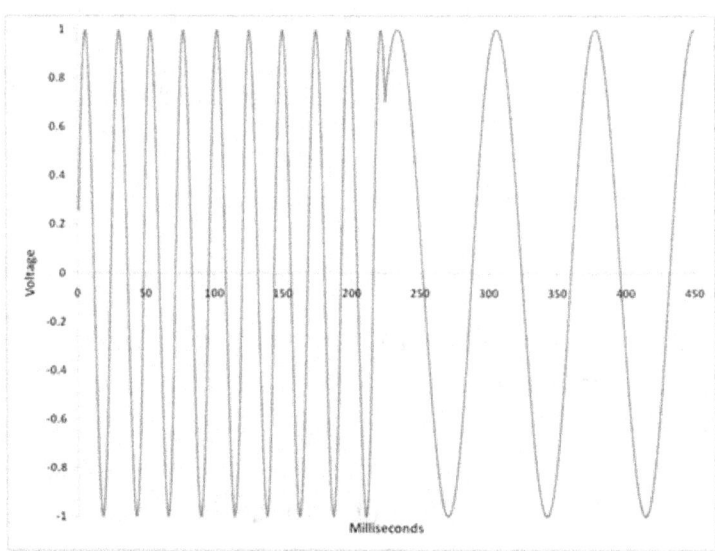

Figure 21. Theoretical example of a cup anemometer response to wind speeds of 32 and 10 m/s.

6.3   Alternative monitoring strategies

We reconfigured the Campbell software to capture the cup anemometer information at 2 Hz. The timeout interval for this rate is 450 ms. Figure 22 is the sonic anemometer information and the cup anemometer information captured at 2 Hz. Even with the 450-ms timeout interval (threshold velocity of 2.01 m/s), there are a large number of values that are NAN (data gaps in the trace for the NRG wind speeds in Fig. 22).

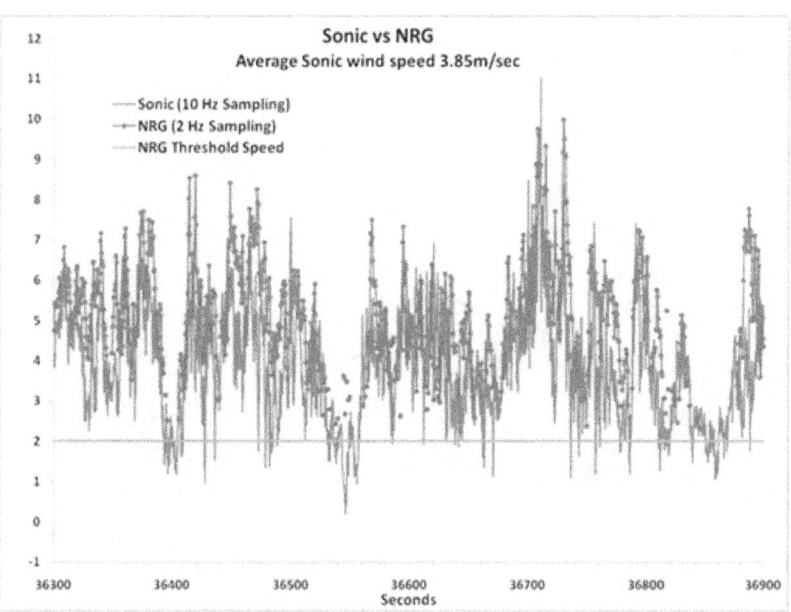

Figure 22. Sonic and cup anemometer winds speed.

If we had used a 5-Hz sampling rate, the number of NAN would have been even greater because the threshold velocity is 4.38 m/s. Increasing the sampling rate reduces the timeout interval and increases the threshold wind speed. While we want a high sampling rate, we also want to measure the wind speed variations at slower wind speeds.

If we operate the Campbell data logger in a mode where we sample the voltage output of the instrument rather than compute the wind speed, we can achieve very high sampling rates. This is possible with the NRG cup anemometer because it uses the Hall Effect (rather than a count) to output a continuous voltage. The sampling rate using the Campbell CR 1000 is 100 Hz. Using this approach requires the development of algorithms to convert the continuous voltage output of the NRG to a frequency and then to convert the frequency information to a wind speed using the NRG calibration equation. Figure 23 represents the voltage output, sampled at 100 Hz, from the NRG cup anemometer. As the wind speed increases, both the frequency and the voltage increase. To convert the voltage output to frequency, we have developed an algorithm that computes the frequency based on the time interval of the

1. Maximum peak-to-peak voltage value (1 to 1 in Fig. 23).
2. Minimum peak-to-peak voltage value (2 to 2 in Fig. 23).
3. Positive to negative voltage value zero crossing (3 to 3 in Fig. 23).
4. Negative to positive voltage value zero crossing (4 to 4 in Fig. 23).

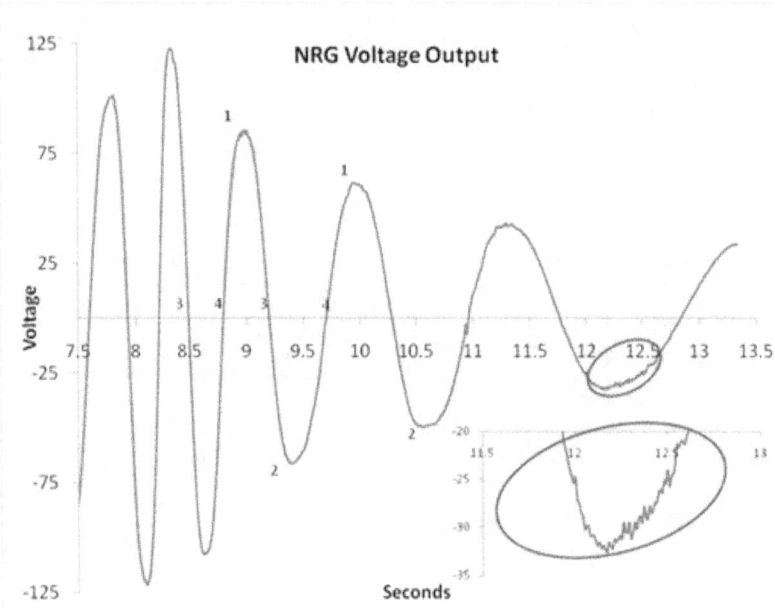

Figure 23. Voltage output from the NRG cup anemometer.

We did not use the peak-to-peak calculation of the frequency because the algorithm interprets the high frequency variations (see insert in Fig. 23) in the voltage signal as changes in frequency. These high frequency values result in unrealistic wind speeds.

To test the algorithm, we collected both NRG and sonic wind speed information simultaneously on two small towers separated by approximately 2 m. The instruments were at a height of approximately 3 m. Turbulence varies both spatially and temporally; and therefore, we do not anticipate the wind speed information and variations in wind speed will be identical for the two instruments. In addition, the NRG will not output a voltage value for wind speeds less than 1 m/s.

Friction and the angular momentum of the spinning cups dampen the response of the cup anemometer to changes in the wind speed. The NRG distance constant (l_o), defined as the length of air passing through the rotor required to reach a new equilibrium rotation rate in response to a change in wind speed, is 3 m; and the moment of inertia is 68×10^{-6} slug-ft². Short-period wind speed fluctuations will be strongly damped if their period, τ_o, is less than

$$\tau_o = l_o/U$$

At higher wind speeds, the value of τ_0 decreases; and more of the short-period fluctuations are captured. The first data presented are for a 10-min period with an average wind speed of 3.96 m/s and a range of wind speed values of 0.95 to 13.1 m/s, corresponding to periods (τ_0) of 0.75, 3.1, and 0.2 s, respectively. Figure 24 shows the eddy velocities for the sonic and NRG instruments and the NRG wind speeds for the 10-min period.

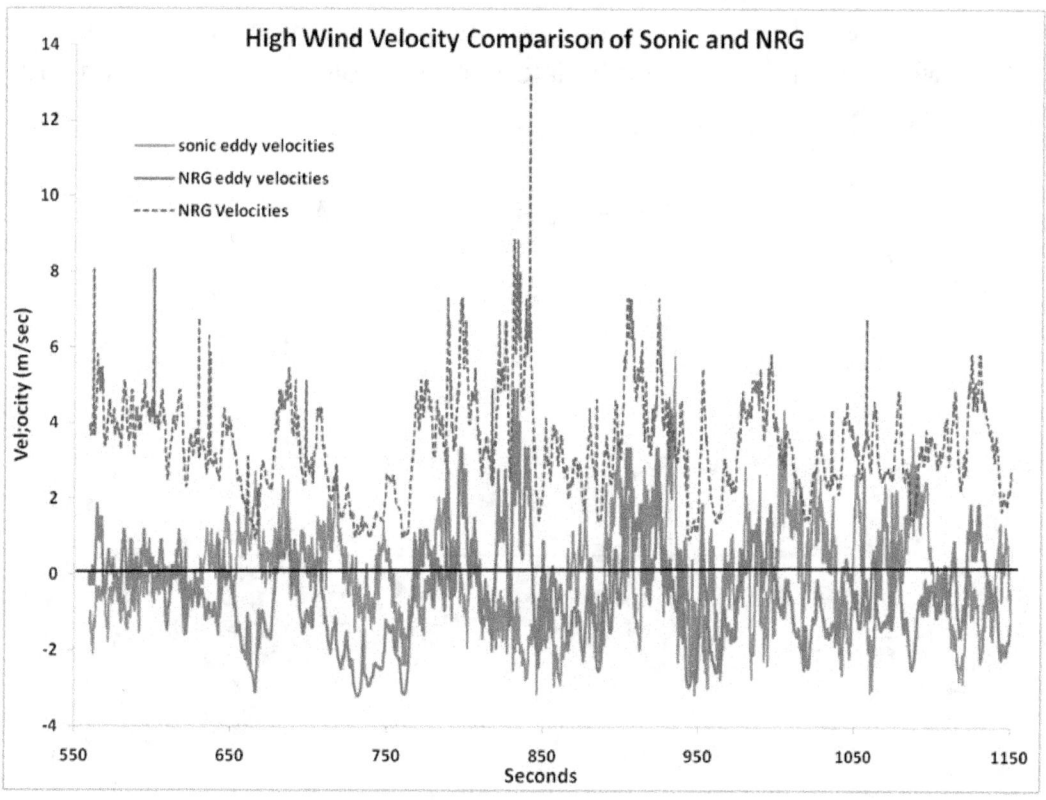

Figure 24. Comparison of the sonic and NRG derived eddy velocities for a period of relatively high wind speeds.

As indicated, we do not anticipate that the sonic and NRG eddy velocities will be identical. When the wind speeds are low, short period velocity fluctuations are filtered out. This is evident in Figure 24 at around 750 s. Around 850 s, the wind speeds are high; and the eddy velocities exhibit higher frequency fluctuations. The gustiness computed from the sonic and NRG eddy velocities for the 10-min period are 0.31 and 0.26, while the turbulent intensities are 0.39 and 0.33. The NRG values for the gustiness and turbulent intensity are much closer to the sonic values as compared to the values given in Table 3 where the sonic gustiness/turbulent intensity was 0.42/1.13 and the NRG gustiness/turbulent intensity was 0.22/0.23. We believe this relatively large disagreement between the sonic and the NRG is a result of the relatively low (10 s) NRG sampling rate. However, when we directly sampled the NRG voltage, we increased the sampling rate from 10 s to 0.01 s, thus capturing more of the high frequency variations in the wind speeds.

It was not possible to collect a continuous 10-min period of low wind speeds using the NRG instrument. The NRG lower wind speed limit is 1 m/s. We did collect approximately a 1.5 min period of values with an average wind speed of 1.8 m/s and a range from 0.61 to 2.62 m/s (Fig. 25). These wind speeds correspond to damping thresholds of 1.63, 4.9, and 1.12 s, respectively.

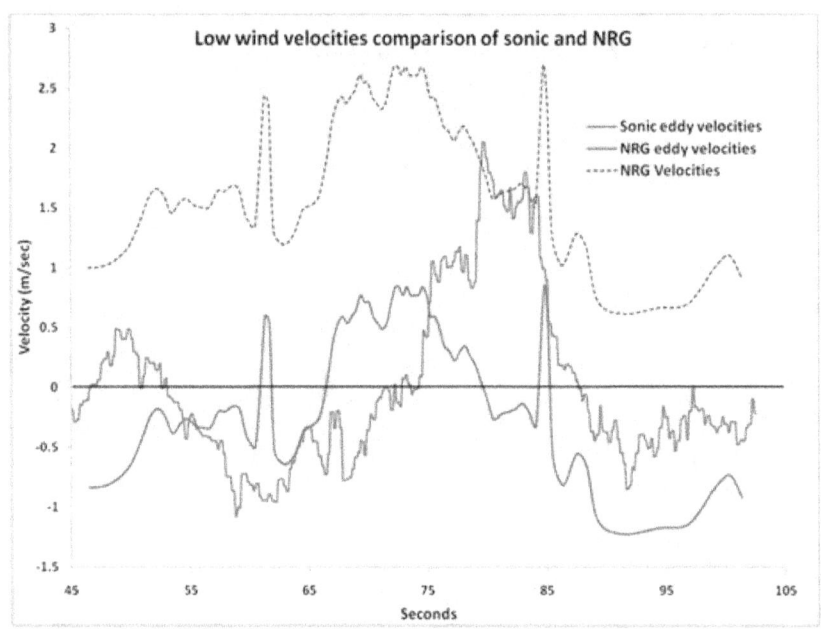

Figure 25. Comparison of the sonic and NRG derived eddy velocities for a period of relatively low wind speeds.

The NRG eddy velocities exhibit more smoothing as anticipated based on the damping threshold as given above. A comparison of the NRG eddy velocity trace reveals more structure in the trace during periods of high wind speeds (65 to 80 s) than during periods of lower wind speeds (90 to 105 s). The sonic gustiness/turbulent intensity for the low wind speed case were 0.38/0.49 while the NRG gustiness/turbulent intensity values were 0.28/0.31. Interestingly, these values are the same order of magnitude as the high wind speed case.

As indicated, our objective is to understand the performance of the VAWT and the HAWT in response to changing wind flow patterns and turbulence in a semi-complex environment. Sonic and cup anemometer measurements will allow us to assess the timing and the frequency and magnitude of turbulence at the turbine demonstration site and to relate the turbulence to turbine efficiency by evaluating degradation of the turbine power curves. Our preliminary investigation into the use of cup anemometers indicated that we could use these instruments to measure eddy velocities, providing that the output voltage was sampled at the highest possible rate. For the CR 1000, this is 100 Hz; and the CR 3000 samples are at an even higher rate of 300 Hz. The cup anemometers, however, do have limitations. The instrument does not output a voltage for wind speeds less than 1 m/s, and the damping of high frequency wind speed fluctuations depends on the instrument distance constant and the wind speed. More damping occurs at lower wind speeds. Monitoring of wind speed and turbine power production will also allow creation of operational envelopes for various weather conditions. This includes the rate and frequency of wind speed ramp-up and ramp-down events, which can create a power management challenge when large numbers of turbines are operating on the grid.

7 Bird and Bat Monitoring Background and Methodology

Bird and bat kills are a very large concern, especially at Fort Drum, which is within the endangered Indiana bat habitat. Large wind turbines are known to kill bats and certain birds. Very little is known about the effects of small turbines on birds and bats even though manufacturers freely claim that it is a non-issue for small turbines.

The majority of studies of bat and bird interactions with wind turbines are for large turbines (BPA 2002; Whittam and Kingsley 2003; NWCC 2010). There is evidence that turbines on shorter towers and with smaller aerial footprints may reduce impact to flying wildlife, but this evidence is not well documented (Andersen 2008). In addition, we have not located any scientific documentation of bird and bat interactions with VAWTs. Though initially the Fort Drum turbines would be limited to daytime operation from April to October, information from monitoring may have allowed night operations in FY13.

FDED planned to perform a pre-construction bird and bat survey in the proposed turbine installation area. After construction, the Fort Drum Environmental Division would have monitored bird and bat kills daily in circular areas cleared around each turbine (Fig. 7). Using stereographic thermal imaging video cameras, ERDC planned to monitor the frequency and behavior of birds and bats near the turbines. Thermal imaging effectively monitors birds and bats near large turbines because it does not require supplementary light sources that may attract wildlife.

7.1 Motivation

Since its first documentation, white-nose syndrome (WNS), a fungal infection that presents as white growth on the muzzle, ears, and wing membrane (Blehert et al. 2009), has ravaged bat populations in the Northeast. Bat colonies in the Northeast have diminished by an average of 75%; and in some instances, colonies have been decimated to near extinction (Dzal et al. 2011; Frick et al. 2010). This devastation of the bat population is alarming. The conservation of bats is not only important with respect to maintaining biodiversity but is also an issue of economic importance. It is estimated that between 660 and 1320 metric tons of insects are now not being eaten by the approximately 1 million bats that have died from WNS (Boyles et al. 2011). This unmitigated growth in the insect population will translate into added crop maintenance costs for insecticide, which have been estimated to be as high as $3.7 billion per year (Boyles et al. 2011).

In addition to the bat population decreases owing to WNS, the expansion of wind energy facilities is an added population pressure for bats and birds. Fort Drum is located in the Eastern US, a region where the annual bat kill rate per commercial-scale turbine ranges from 15 to 39 bats per turbine, which is significantly higher compared to other regions of the country. The Midwest has the next highest kill rate at 8 bats per turbine (Arnett et al. 2008). While the impact on birds is significantly smaller than for bats, with annual kill rates much less than 1 bird per turbine for the Eastern US (Barclay et al. 2007), the reduction or elimination of these bird kills is still a concern for this study. However, the main focus remains primarily on bat interactions with wind turbines because bats' low reproductive rates make population recovery more difficult and the reduction of kill rates much more urgent (Frick et al. 2010).

The majority of studies of bat and bird interactions with wind energy facilities are for utility-scale turbines (> 1 MW) with blade diameters ranging from 40 to 90 m with rotor swept areas of 1250–6400 m² (Arnett et al. 2008; Barclay et al. 2007; Baerwald et al. 2008; Cryan and Brown 2007; Cryan and Barclay 2009; Horn et al. 2008a; Horn et al. 2008b; Smallwood 2007). The small Fort Drum turbine would have had blade diameters in the range of 2 to 5 m with rotor swept areas of 3 to 20 m². There is evidence that turbines on shorter towers and with smaller aerial footprints may reduce impact to flying wildlife (Barclay et al. 2007), but bird and bat interactions with smaller wind turbines (power output of about 5 kW) have not been well documented (Andersen 2008). In fact, the details of bat and bird interactions with wind turbines are not well understood for turbines of any size (Boyles et al. 2011). As we have not located any scientific documentation of bird and bat interactions with vertical-axis turbines, Army Environmental and Energy Managers would benefit from information regarding the behavior and prevalence of birds and bats in proximity to VAWTs.

7.2   Methodology

Plans were to have a jointly-executed bird and bat kill monitoring program at Fort Drum, with one methodology executed by the FDED and the other by CRREL. The FDED would have monitored bird and bat kills within circular areas with radii equivalent to 1.5 tower heights cleared around each turbine. FDED would have inspected these cleared areas daily for bird or bat carcasses. With stereographic thermal imaging of the rotor-swept areas using two FLIR LWIR (long-wave infrared - 7.5 to 13 μm) video cameras framing at a maximum rate of 30 Hz, CRREL would have complimented the bird and bat kills work by monitoring the prevalence and flight behavior of birds and bats near the turbines. Thermal imaging has been widely used for visual tracking and monitoring of bird and bat populations near large turbines (Horn et al. 2008a; Hristov et al. 2010; Gauthreaux et al. 2006), an attractive approach because of the ability to detect and track the animals at night without requiring a supplementary light source.

Studies of bats around large turbines suggest that migratory patterns may affect bat mortality rates, and weather conditions may lead to flight behavior that puts the animals at an additional risk of collision (Cryan and Brown 2007). To capture migratory and weather effects on bat-turbine interaction, we would have acquired video over 10-day periods during August to November and March to May, migration periods for tree-roosting migratory bats, which are the type of bats that are disproportionately affected by wind turbines (Arnett et al. 2008; Barclay et al. 2007; Baerwald et al. 2008; Cryan and Brown 2007; Cryan and Barclay 2009). Our approach would have been similar to that of Horn et al. (2008b), who monitored bats at the Maple Ridge Wind Farm in Lowville, NY, which is less than 25 miles from Fort Drum. Bat monitoring would begin 10 min after sundown and continue for 2 hr to capture the period that the animals are most likely to be active around the turbines (Horn et al. 2008b). Custom tracking software would identify flight trajectories in each of the two camera views. A direct linear transformation (DLT) (Shapiro 1978) would map these 2D flight paths into 3-dimensional space, determining the animal's altitude, 3D flight path, and proximity to the wind turbines.

Assuming a tower height of 25 m, we would have placed the IR cameras 10 m from the tower base and 27 m from the turbine hub (Fig. 26). Each camera has view angles of 25° (horizontal) and 18.8° (vertical), and the field-of-view dimensions at the turbine hub are 12.2 (horizontal) × 8.9 m (vertical). We would have monitored each turbine individually during each

10-day period. For example, we would have position the cameras to monitor only the VAWT for 5 days. At the conclusion of these 5 days, we would have repositioned the cameras to monitor the HAWT for the next 5 days. For each day, we would have recorded video starting at 20 min past sunset and ending at 3 hr, 40 min after sunset. The video acquisition would synchronize the two cameras because errors in the time registration of the video pair would lead to errors in the 3-dimensional reconstruction. The video images from each camera would have streamed via Ethernet to an onsite computer at a maximum rate of 30 Hz and stored on an internal hard drive for post-processing offsite at the conclusion of the 10-day monitoring session.

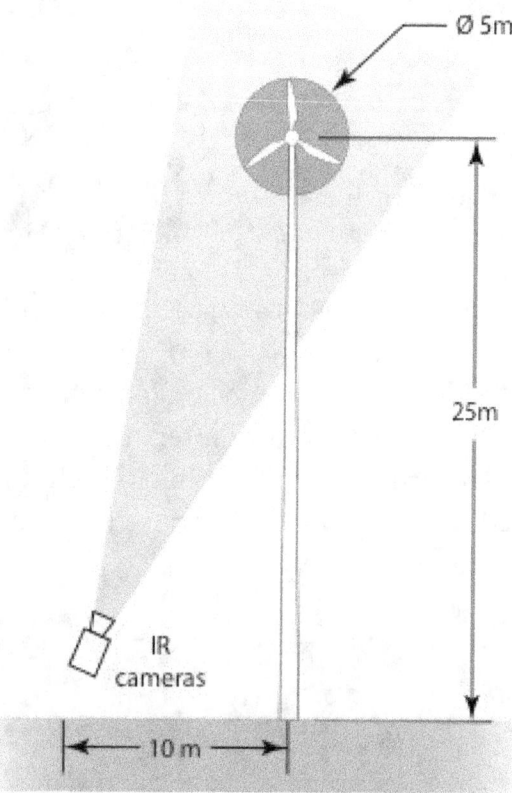

Figure 26. IR camera configuration.

7.3 Preliminary measurements

We conducted a proof-of-concept demonstration at a central New Hampshire location where at least two individuals of the little brown bat (*Myotis lucifugus*) have been sighted. Only one IR camera was available for this preliminary measurement. Therefore, we can only demonstrate the ability to identify and track the bats using thermal imaging and cannot demonstrate 3-dimensional reconstruction of the flight paths.

The test site was a grassy field with tall trees lining the boundary. We positioned the camera so its view was unobstructed by foliage and provided a high thermal contrast background with a bat's body, as seen in Figure 27 (left). For the weather conditions during the acquisition (Table 4), there was poor thermal contrast between the foliage and bats. We started the acquisition at 10 min after sunset (2039 hr) and continued for 2 hr (2239 hr).

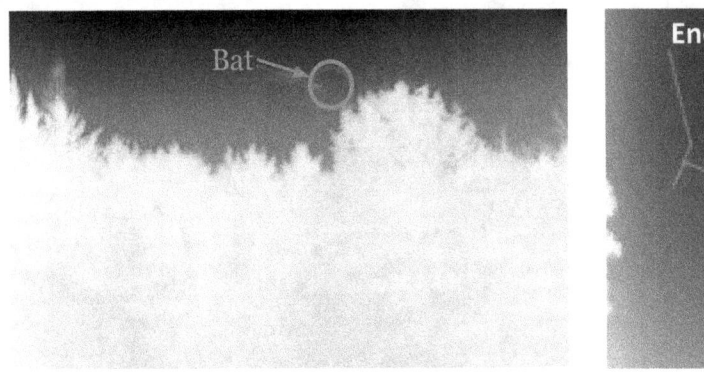 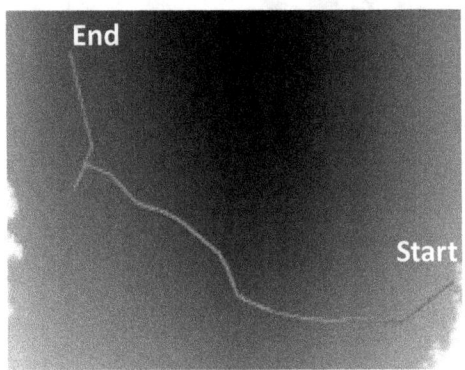

Figure 27. Example IR image and flight tracking results.

We then parsed the video into shorter segments where a bat was visible in a frame and tracked it within this short interval. A custom MATLAB program was written to record the bat's location in each frame using mouse clicks in the image. Figure 27 (right) illustrates the tracking results of one of these shorter segments.

8   Icing Monitoring

The impact of icing from freezing rain and rime icing on wind turbines is known to be a significant problem for many large turbines (Morgan et al. 1998; Seifert et al. 2003; Frohboese and Anders 2007; Jasinski et al. 1997; Tammelin and Seifert 2001). Icing decreases operating efficiency by modifying blade aerodynamics. Reduced power output from decreased efficiency or stoppage because of icing impacts turbine economics. Icing stresses hardware when blades become imbalanced from the differential shedding of ice. When the blades are rotating and the ice is beginning to melt and shake loose, bits of thrown ice can cause hazards to nearby personnel and materiel. Large turbines in Europe are documented to have thrown several-kilogram pieces of shed ice hundreds of meters—clearly a hazard.

Small turbine manufacturers generally state that icing is not a problem for them. Some state that their turbines do not ice, and others claim that icing decreases their blade rotation speed sufficiently to not cause an ice-throw hazard. Unfortunately, formal studies are not available that demonstrate whether icing is a problem for small turbines.

ERDC and Fort Drum personnel had planned to monitor icing on the two demonstration turbines using ice detectors, video cameras, and a weather station. Fort Drum personnel would have monitored the area around each turbine base for ice thrown from the blades. A CCD camera fed to CRREL would have monitored icing on the blades during daylight hours because, to reduce bat and bird encounters with the blades, lighting at night was not allowed at the site.

A freezing rain detector, a Goodrich (Rosemount) 872E3, would indicate when icing was occurring near ground level; but it would not indicate if in-cloud icing was occurring on the rotors when icing was not occurring at ground level (Fig. 28). The 872E3 detector is similar to ice detectors used by the DoD, FAA, and National Weather Service at Automated Surface Observing Systems (ASOS) stations to monitor icing at 600 airports nationwide (Ryerson and Ramsay 2007). The ice detector only indicates when ice is actually accumulating. It does not indicate how long ice resides on objects after actual icing conditions have ceased.

Several sources concurrently would determine the resident time of ice on the blades: (a) the ice detector indicates when icing is occurring; (b) if there is wind, the unheated anemometers should also be icing and will either freeze to a stop or slow; (c) the standard deviation of wind speed decreases during icing because the response time of the iced anemometers increases; and (d) the unheated wind direction vanes cease to function, and their standard deviation decreases from resident ice. The camera would indicate that ice is on the turbines during daylight, and personnel visiting the site would corroborate that ice is on the turbine blades.

Once ERDC detected an icing event and its magnitude, it had planned to determine the effects of ice- and snow-laden blades on turbine efficiency through changes in the turbine power curves and to indicate the duration and magnitude of power losses caused by ice and snow.

Figure 28. Goodrich 872E3 ice detector.

9   Status and Plans

FY11 was the first year of this project. The initial plan was to obtain turbine permitting in FY11, to install the turbines, and to conduct research on the turbines at Fort Drum. The latter would have included radar measurements, seismic-acoustic measurements, and initial bird and bat and turbulence measurements. In addition, weather and power output measurements would have begun in FY11. The program was to continue monitoring the turbines, turbulence, and bird and bat kills into 2012 and to conduct icing effects studies, given appropriate weather conditions, during the winter of FY12.

We started the project in the first quarter of fiscal year 2011 before funding arrived because we anticipated needing additional time for permitting and to select and purchase the turbines. However, permitting proved to be a longer process than anticipated. Even though funding arrived in March 2011, the Principal Investigator and the ITTP program agreed that turbine purchase, installation, and scientific work conducted after the turbines were operational would be delayed one-year until late FY12. The Principal Investigator advised the ITTP Program that it would be prudent to not purchase the turbines before completing permitting because if the turbines were not approved in the permitting process, the ITTP program would own two turbines that could not be installed. Approximately $330K out of $475K awarded in 2011 was returned to the ITTP program in May 2011 for this reason. An ITTP budget cut occurred in FY12, so funding for this project was not continued because it had the largest budget.

The new schedule for the program would have been as follows: FY11 was consumed with turbine and site selection, permitting, and developing scientific methodology; in FY12, we would have purchased and installed the turbines and started initial scientific work. We would have completed scientific work and monitoring in FY13, culminating with a draft small wind turbine UFC.

9.1   FY11 accomplishments

We accomplished the following tasks in Year 1 of the project:

1. Completed permitting. ERDC was informed in August FY11 that FAA permitting had been approved and that environmental permitting had also been approved but that paperwork would not arrive until early FY12.
2. Selected turbines and a demonstration site on Fort Drum.
3. Developed scientific methodologies for measuring and documenting radar effects, seismic-acoustic effects, turbulence effects, bird and bat kills, and icing effects and for determining turbine efficiency.
4. Developed plans for collaborating with the Fort Drum Environmental Division for monitoring bird and bat kills.

9.2 FY12 plans

Plans for FY12 would have begun in March 2012:

1. Purchase turbines.
2. Prepare site and install turbines.
3. Order final miscellaneous hardware and instrumentation.
4. Install weather and turbulence equipment.
5. Conduct radar, seismic-acoustic, initial turbulence, and initial bird and bat measurements.
6. Contract with Clarkson University for FY13 monitoring assistance between October FY13 and April FY13 (no ITTP funding available for ERDC during that period).
7. Begin draft UFC.

9.3 FY13 plans

1. Continue bird and bat monitoring and turbulence monitoring.
2. Begin and complete icing monitoring.
3. Complete all turbulence, radar, seismic-acoustic, bird and bat, icing, and power output monitoring.
4. Address effects of many small turbines on facilities.
5. Conduct and demonstrate economic analysis.
6. Complete final report and draft UFC.

10 Army Deliverables and Benefits

Currently, there is little information available within DoD to guide facility and installation managers regarding the planning and construction of small wind turbines. Government, academia, nor industry have provided information regarding the compatibility of small wind turbines with military facilities and their activities. There is no guidance on the effect of small HAWT and VAWT on radar and on bird and bat kills nor on their seismic and acoustic effects. Additionally, there is little guidance on the effects of turbulence and icing on small turbine efficiency, safety, and integrity. There is no Unified Facilities Criteria document that addresses the planning and construction of small wind turbines on Army installations. Even the 2007 UFC for sustainable buildings has no link or reference to wind turbines. Our project would have delivered a UFC that addresses all issues for small wind turbines planned or owned by the Army.

Topics that this UFC would have included:

1. The formula for wind power and annual wind energy output with an explanation of terms and their importance in selecting and siting a wind turbine or turbines.
2. The importance of proper turbine siting to eliminate turbulence and to provide maximum access to prevailing and secondary winds without impact from potential obstructions. This includes siting the turbine so that the tower hub height is 30-ft above the tallest object within a 500-ft radius plus the radius of the turbine blades.
3. The effects of small HAWT and VAWT on radar and on bird and bat kill, the effects of icing and turbulence and their seismic-acoustic signature.
4. The sources of models and guidance to predict shadow flicker, noise, electromagnetic effects, radio frequency interference (RFI) and microwave interference, annual energy production, and long term maintenance cost projections.
5. Alternative contracting methods that facilities can use to install a wind turbine, such as Energy Savings Performance Contracts (ESPC), Utility Energy Service Contracts (UESC), and Power Purchase Agreements (PPA) with a simple lease.
6. Examples of lessons learned from turbines already installed and operating on military facilities.

7. Permitting guidance.

Additional technology transfer would have occurred via ERDC reports, DoD and other conference presentations, and journal articles. All documents and an Executive Summary would have recommended how to implement small wind turbines Army-wide. A workshop at a DoD energy conference, such as the annual ERDC-CERL Energy Workshop, may also have been a deliverable.

11 Conclusions

Below is a summary of the project tasks that were completed:

- Conducted an EA with a finding of No Significant Impact and determined that the project did not require a Clean Air Act General Conformity Determination nor need an Environmental Impact Statement.
- Published a public notice that remained open from 3 July 2011 to 1 August 2011.
- Received permission from the FDED in August 2011 to proceed with the project.
- Obtained an FAA permit.
- Obtained a USFWS permit.
- Received project approval from the Department of Defense Energy Siting Clearinghouse.
- Evaluated turbine demonstration sites, and Fort Drum selected and approved the optimal site.
- Selected for purchase two turbines, one 7.7-kW horizontal-axis wind turbine and one 2.9-kW vertical-axis wind turbine, based on the criteria set forth by Fort Drum and CRREL.
- Conducted laboratory tests to collect a radar backscatter measurement from the full-scale wind turbines at Fort Drum.
- Developed a methodology to measure the seismic acoustic signature and used past studies with similarities to the Fort Drum study as background information for this project.
- Developed a methodology to measure and to understand the performance of small turbines as they respond to changing wind flow patterns and turbulence in a semi-complex environment.
- Developed a methodology to monitor bird and bat kills at the turbine site and developed a setup using stereographic thermal imaging video cameras to monitor the frequency and behavior of birds and bats near the turbines. Executed preliminary measurements with the thermal cameras.
- Developed a methodology to monitor icing events on the turbines.

This project supported policies such as *The Army Strategy for the Environment* (2004) and the *Army Energy Strategy for Installations*. Had execution to completion been enabled, this project would have aided energy managers, through a Unified Facilities Criteria and other reports and workshops, in deciding whether to adopt small wind turbines. It would have reduced wasteful spending on wind installations not compatible with DoD facilities and operations.

References

Albert, D. G., and F. E. Perron. 2006. *Wind Turbine Induced Seismic Noise—Preliminary Findings*. Unpublished contract report.

Andersen, K. W. 2008. *A Study of the Potential Effects of a Small Wind Turbine on Bird and Bat Mortality at Tom Ridge Environmental Center, Erie, Pennsylvania*. Unpublished report. Erie, PA: Gannon University.

Arnett, E. B.,W. K. Brown, W. P. Erickson, J. K. Fiedler, B. L. Hamilton, T. H. Henry, A. Jain, G. D. Johnson, J. Kerns, R. R. Koford, C. P. Nicholson, T. J. O'Connell, M. D. Piorkowski, and R. D. Tankersley, Jr. 2008. Patterns of Bat Fatalities at Wind Energy Facilities in North America. *Journal of Wildlife Management* 72:61–78.

Arnfield, A. J. 1984. Simulating radiative energy budgets within the urban canopy layer. *Modeling and Simulation* 15:227–233.

AWEA (American Wind Energy Association). 2009a. *AWEA Small Wind Turbine Performance and Safety Standard*. AWEA Standard AWEA 9.1-2009. Washington, DC: American Wind Energy Association.

Baerwald, E. F., G. H. D'Amours, B. J. Klug, and R. M. R. Barclay. 2008. Barotrauma is a significant cause of bat fatalities at wind turbines. *Current biology* 18:R695–R696.

Barclay, R. M. R., E. F. Baerwald, and J. C. Gruver. 2007. Variation in bat and bird fatalities at wind energy facilities: assessing the effects of rotor size and tower height. *Canadian Journal of Zoology* 85:381–387.

Belote, D. 2011. The 2011 NDAA and DoD's Energy Siting Clearinghouse. Washington, DC: DoD Energy Siting Clearinghouse, Office of the Secretary of Defense. http://www1.eere.energy.gov/femp/pdfs/fupwg_fall11_belote.pdf.

Blehert, D. S., A. C. Hicks, M. Behr, C. U. Meteyer, B. M. Berlowski-Zier, E. L. Buckles, J. T. H. Coleman, S. R. Darling, A. Gargas, R. Niver, J. C. Okoniewski, R. J. Rudd, and W. B. Stone. 2009. Bat white-nose syndrome: an emerging fungal pathogen? *Science* 323(5911):227.

Boyles, J. G., P. M. Cryan, G. F. McCracken, and T. H. Kunz. 2011. Conservation: Economic importance of bats in agriculture. *Science* 332:41–42.

BPA (Bonneville Power Administration). 2002. *Synthesis and Comparison of Baseline Avian and Bat Use, Raptor Nesting and Mortality Information from Proposed and Existing Wind Developments*. Portland, OR: Bonneville Power Administration.

Colby, W. D., R. D. Dobie, G. Leventhall, D. M. Lipscomb, R. J. McCunney, M. T. Seilo, and B. Søndergaard. 2009. *Wind Turbine Sound and Health Effects, An Expert Panel Review*. American Wind Energy Association and Canadian Wind Energy Association.

Cryan, P. M., and R. M. R. Barclay. 2009. Causes of Bat Fatalities at Wind Turbines: Hypotheses and Predictions. *Journal of Mammalogy* 90:1330–1340.

Cryan, P., and A. Brown. 2007. Migration of bats past a remote island offers clues toward the problem of bat fatalities at wind turbines. *Biological Conservation* 139:1–11.

DoD (US Department of Defense). 2006. *The Effect of Windmill Farms on Military Readiness*. Report to the Congressional Defense Committees. Office of the Director of Defense Research and Engineering.

Dzal, Y., L. P. McGuire, N. Veselka, M. B. Fenton, M.B. 2011. Going, going, gone: the impact of white-nose syndrome on the summer activity of the little brown bat (Myotis lucifugus). *Biology letters* 7:392–394.

FAA (Federal Aviation Administration). 2007. *Change 2 to Obstruction Marking and Lighting*. Advisory Circular 70/7460-1 K Change 2. US Department of Transportation, Federal Aviation Administration.

FDED (Fort Drum Environmental Division). 2011. *Fort Drum Wind Power: Offsetting One Environmental Issue Without Creating Another*. Unpublished 15-slide briefing, Jason Wagner PW ENV, 4 February (see Appendix 4).

Frick, W. F., J. F. Pollock, A. C. Hicks, K. E. Langwig, D. S. Reynolds, G. G. Turner, C. M. Butchkoski, and T. H. Kunz. 2010. An emerging disease causes regional population collapse of a common North American bat species. *Science* 329:679–682.

Frohboese, P., and A. Anders. 2007. Effects of icing on wind turbine fatigue loads. *Journal of Physics, Conference Series* 75:012061.

Gauthreaux, S. A. and J. W. Livingston. 2006. Monitoring bird migration with a fixed-beam radar and a thermal-imaging camera. *Journal of Field Ornithology* 77:319–328.

Gipe, P. 2004. *Wind Power*. White River Junction, VT: Chelsea Green Publishing Company.

Graettinger, C., S. Garcia, J. Siviy, R. Schenk, and P. Van Syckle. 2002. *Using the technology readiness levels scale to support technology management in the DoD's ATD/STO environments*. Special Report CMU/SEI-2002-SR-027. Pittsburgh, PA: Software Engineering Institute, Carnegie Mellon University.

Horn, J. W., E. B. Arnett, and T. H.Kunz. 2008a. Behavioral responses of bats to operating wind turbines. *The Journal of Wildlife Management* 72:123–132.

Horn, J. W., E. B. Arnett, M. Jensen, T. H. Kunz. 2008b. *Testing the effectiveness of an experimental acoustic bat deterrent at the Maple Ridge wind farm*. Boston, MA: Boston University Center for Ecology and Conservation Biology.

Hristov, N. I., M. Betke, D. E. H. Theriault, A. Bagchi, and T. H. Kunz, T.H. 2010. Seasonal variation in colony size of Brazilian free-tailed bats at Carlsbad Cavern based on thermal imaging. *Journal of Mammalogy* 91:183–192.

Hubbard, H. H., and K. P. Shepherd. 1991. Aeroacoustics of large wind turbines. *Journal of the Acoustical Society of America* 89:2495–2508.

IEC (International Electrotechnical Commission). 2006. Wind Turbine generator systems—Part 11: Acoustic noise measurement techniques. International Standard IEC 61400-11, Edition 2.1. Geneva, Switzerland: International Electrotechnical Commission.

Jasinksi, W. J., S .C. Noe, M. S. Seig, and M. B. Bragg. 1997. *Wind turbine performance under icing conditions.* AIAA 97-0977. AIAA, 35th Aerospace Sciences Meeting and Exhibit, 6–9 January, Reno, NV. American Institute of Aeronautics and Astronautics.

Juhl, D. 2011. *Permitting Roles and Responsibilities.* AWEA Small and Community Wind Conference and Exhibition, 15–17 September, Des Moines, IA.

Manwell, J. F, J. G. McGowan, and A. L. Rogers. 2002. *Wind Energy Explained—Theory, Design and Application.* Chichester, UK: John Wiley & Sons Ltd.

Morgan, C., E. Bossanyi, and H. Seifert. 1998. *Assessment of safety risks arising from wind turbine icing.* BOREAS IV, 31 March–2 April, Hetta, Finland.

NDAA (National Defense Authorization Act). 2011. *Section 358 Study Of Effects Of New Construction Of Obstructions On Military Installations And Operations. National Defense Authorization Act.*

Niver, R. 2009. *Biological Opinion on the Proposed Activities on the Fort Drum Military Installation (2009–2011) for The Federally-Endangered Indiana Bat (Myotis sodalis).* Cortland, NY: US Fish and Wildlife Service.

NOAA (National Oceanic and Atmospheric Administration). 2011. *How Rotating Wind Turbine Blades Impact the NEXRAD Doppler Weather Radar.* Norman, OK: NOAA's National Weather Service Radar Operations Center. http://www.roc.noaa.gov/wsr88d/windfarm/turbinesimpacton.aspx?wid=dev.

NWCC (National Wind Coordinating Collaborative). 2010. *Wind Turbine Interactions with Birds, Bats, and their Habitats: A Summary of Research Results and Priority Questions.* Washington, DC: National Wind Coordinating Collaborative.

Oke, T. R. 1976. The distinction between canopy and boundary-layer heat islands. *Atmosphere* 4:268–277.

Oke, T. R. 1979. Advectively-assisted evapotranspiration from irrigated urban vegetation. *Boundary-Layer Meteorology* 17:167–173.

Pedersen, E., and K. P. Waye. 2004. Perception and annoyance due to wind turbine noise—a dose-response relationship. *Journal of the Acoustical Society of America* 116:3460–3470.

Rowley, S. 2009. Personal communication with Charles Ryerson. Fort Drum, NY.

Ryerson, C., and A. Ramsay. 2007. Quantitative Ice Accretion Information from the Automated Surface Observing System (ASOS). *Journal of Applied Meteorology and Climatology* 46:1423–1437.

Sagrillo, M. 2011. *Permitting Small Wind Systems, Wisconsin's Small Wind Model Zoning Ordinance.* Presentation at AWEA Small and Community Wind Conference and Exhibition, 15 September, Des Moines, IA.

Seifert, G. 2006. *Wind Radar Interference.* Briefing Idaho National Lab.

Seifert, H., A. Westerhellweg, and J. Kröning. 2003. *Risk analysis of ice throw from wind turbines.* BOREAS VI, 9–11 April, Pyhä, Finland.

Shapiro, R. 1978. Direct Linear Transformation Method for Three-Dimensional Cinematography. *Research Quarterly* 49:197–205.

Smallwood, K. S. 2007. Estimating Wind Turbine-Caused Bird Mortality. *Journal of Wildlife Management* 71:2781–2791.

Tammelin, B., and H. Seifert. 2001. *Large wind turbines go into cold climate regions.* European Wind Energy Conference 2001, Copenhagen, 2–6 July, Copenhagen, Denmark.

Toth, M., E. Jones, D. Pittman, and D. Soloman. 2011. DOW Radar Observations of Wind Farms. *Bulletin of the American Meteorological Society*, August 2011:987–995.

US Army. 2011. *Environmental Assessment for Conducting a Study of Small Wind Turbines on Fort Drum, New York.* Fort Drum, NY: Directorate of Public Works Environmental Division Natural Resources Branch.

Weather Underground, Inc. 2013. *Weather Underground.* http://www.wunderground.com/.

Whittam, B., and A. Kingsley. 2003. Shades of Green: A Bird's Eye View of Wind Energy. *BirdWatch Canada* 23:4–7.

Appendix 1: Environmental Assessment

**FOR
CONDUCTING A STUDY OF SMALL WIND TURBINES ON
FORT DRUM, NEW YORK**

June 2011

Prepared By:

Directorate of Public Works
Environmental Division
Natural Resources Branch
Fort Drum, New York 13602

This document is to be referenced with the following citation:

> US Army, 2011. *Environmental Assessment for Conducting a Study of Small Wind Turbines on Fort Drum, New York.* Prepared by the Directorate of Public Works Environmental Division Natural Resources Branch, Fort Drum, NY. June 2011.

ACRONYMS AND ABBREVIATIONS

ACHP	Advisory Council on Historic Preservation
AR	Army Regulation
BO	Biological Opinion
BCA	Bat Conservation Area
CEQ	Council on Environmental Quality
CFR	Code of Federal Regulations
CX	categorical exclusion
dB	Decibels (noise unit of measure)
dBA	A-Weighted Decibels
DOD	Department of Defense
EA	Environmental Assessment
EIS	Environmental Impact Statement
EROI	Energy Return on Investment
ESA	Endangered Species Act
FNSI	Finding of No Significant Impact
FWS	Fish and Wildlife Service
ha	Hectares
NEPA	National Environmental Policy Act
NHPA	National Historic Preservation Act
NYS	New York State
NPDES	National Pollutant Discharge Elimination System
RTLP	Range Training Land Plan
SHPO	State Historic Preservation Office
SME	Subject Matter Experts
US	United States
USFWS	United States Fish and Wildlife Service
USC	United States Code
USGS	United States Geological Survey
VEC	Valued Environmental Components

ENVIRONMENTAL ASSESSMENT
FOR CONDUCTING A STUDY OF SMALL WIND TURBINES ON
FORT DRUM, NEW YORK

TABLE OF CONTENTS

APPENDICIES

**ENVIRONMENTAL ASSESSMENT
FOR CONDUCTING A STUDY OF SMALL WIND TURBINES ON
FORT DRUM, NEW YORK**

1.0 PURPOSE, NEED, AND SCOPE

1.1 Introduction

Fort Drum currently has no wind generation capacity developed on the installation and wind power is one of the most viable alternative energy options in northern New York. However, large wind turbines have be shown to have significant impacts to birds and bats as well as to airfield radar equipment and have therefore been excluded as a suitable alternative energy option for the installation at this time. Small wind systems conversely have had only limited research done on their affects to birds and bats and airfield radar arrays, and so far have demonstrated none of the significant negative effects of large wind turbines. Therefore, Fort Drum has partnered with EDRC-CRREL to explore the possibilities that small wind turbines could provide a valuable alternative energy option for Fort Drum and other Army installations.

This EA has been developed in accordance with NEPA; the regulations issued by the Council on Environmental Quality (CEQ), 40 CFR Parts 1505,1508; and the Army's implementing procedures in 32 CFR Part 651, *Environmental Analysis of Army Actions*. A specific requirement for this EA is an appraisal of impacts of the proposed project, including a determination of a Finding of No Significant Impact (FNSI) or a Notice of Intent (NOI) to prepare an Environmental Impact Statement (EIS).

32 CRF 651 is the regulation used to establish policy, procedures, and responsibilities for assessing environmental effects of Army actions. Section 651.29 screening criteria section (b) (12), specifically states that Establishes a precedent (or makes decisions in principle) for future or subsequent actions that are reasonably likely to have a future significant effect and (c) (1) if the proposed action would adversely affect ...threatened or endanger species or there designated habitat requires preparation of an Environmental Assessment (EA). The proposed study and the associated EA are required to document the need and monitoring requirements of the small wind study on Fort Drum.

1.2 Purpose and Need

The purpose of the Proposed Action is to identify and evaluate environmental consequences of studying two types of small wind turbines to determine the viability of these types of systems for use on Fort Drum and other Army Installations.

The Proposed Action is restricted to analysis of the installation, biological monitoring and power generated to cost ratios for two small wind turbines on Fort Drum. The environmental effects of implementing the proposed action on Fort Drum are the focus of this analysis.

1.3 Scope

This EA addresses potential impacts to environmental resources, such as vegetation, wildlife, threatened and endangered species, soils, climate, air, noise, wetlands and water resources, recreation, socioeconomics and cultural resources. The EA was prepared utilizing a systematic, interdisciplinary approach integrating the natural and social sciences with planning and decision-making. This EA serves as a decision-making tool for the siting of the study site for the Proposed Action.

1.4 Project Location & Description

Criteria for the project location requires a site where the greatest amount of data could be gathered in close proximity to the radar tower array, easily accessible, near the power grid, in an open space that would accommodate both towers and the area defined by necessary standoff distances for monitoring bat, bird, and ice concerns.

2.0 DESCRIPTION OF THE PROPOSED ACTION AND ALTERNATIVES

2.1 Description Of The Proposed Action:

Fort Drum currently has no wind generation capacity developed on the installation and wind power is one of the most viable alternative energy options in northern New York. However, large wind turbines have be shown to have significant impacts to birds and bats as well as to airfield radar equipment and have therefore been excluded as a suitable alternative energy option for the installation at this time. Small wind systems conversely have had only limited research done on their affects to birds and bats and airfield radar arrays, and so far have demonstrated none of the significant negative effects of large wind turbines. Therefore, Fort Drum has partnered with EDRC-CRREL to explore the possibilities that small wind turbines could provide a valuable alternative energy option for Fort Drum and other Army installations.

The Army proposes to support the construction of two study wind turbines, one vertical axis and one horizontal axis, and to study the operation of these wind turbines to determine feasibility of employing these types of systems at Fort Drum. Until this study is completed, and based on findings, the turbines will be shut down during periods of potential bats strikes to eliminate any chance of an Endangered Species Act (ESA) violation. The species impact component of the study can only be done after the formal consultation with USFWS and the biological Opinion for the next three years is completed and approved. This EA is only to explore the locations and placement of the study wind turbines and the alternatives for that location.

2.2   Background

It is important to note that implementing the Proposed Action would not alter the essential nature of Fort Drum, which would remain as a military installation on which Soldiers train, work, and live, and on which there are facilities to support those activities.

Fort Drum, the 10th Mountain Division of the United States Department of the Army, is located approximately 10 miles northeast of the City of Watertown, New York and 15 miles to the east of Lake Ontario (Figure 2.1, Fort Drum, New York Area Map). The Installation is approximately 80 miles north-northeast of Syracuse, New York and can be accessed by the major highway routes of U.S. Interstate 81, New York Highways 3, 11, 26 and 342, in addition to several smaller county and Installation-maintained roads.

Fort Drum has 41 ranges and over 75,000 acres of maneuver training lands to train approximately 80,000 troops per year. The Installation occupies approximately 107,265 acres of land within the Great Lakes drainage basin. Fort Drum lies within Jefferson and Lewis Counties and is adjacent to St. Lawrence County, New York. The northeastern portion of the installation includes the western portion of the Adirondack Mountains of the New York State Park.

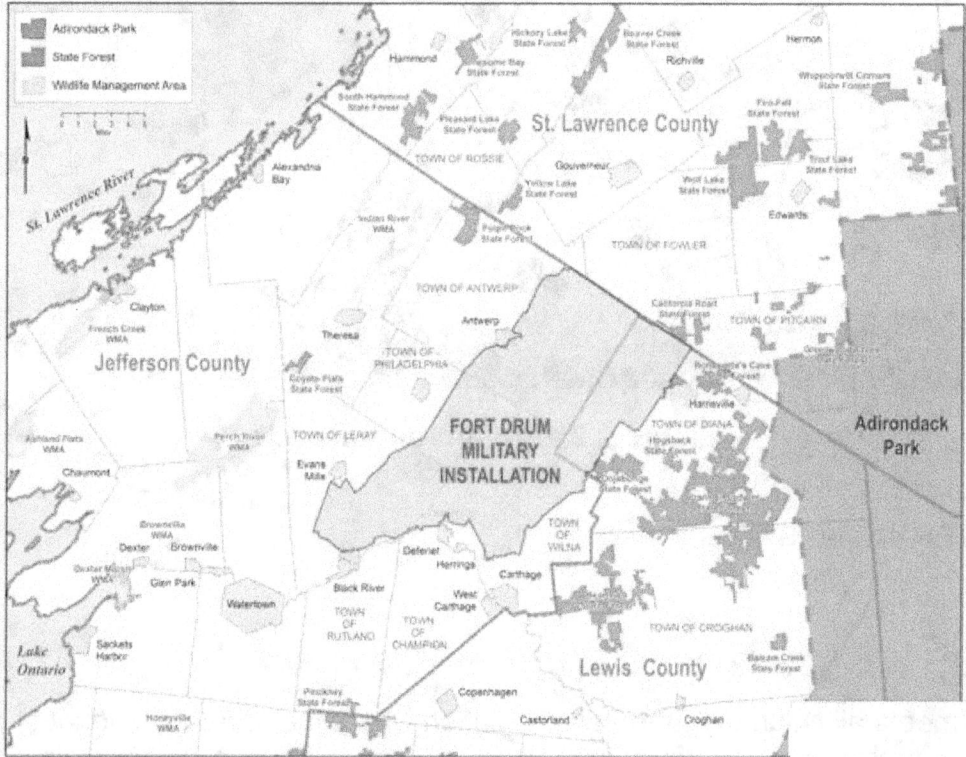

Site location is shown on Land-cover map with Natural Resources Management Units (NRMU) classifications for the proposed location shown on Figure 2.2 below There are no wetland or cultural resources associated with the location.

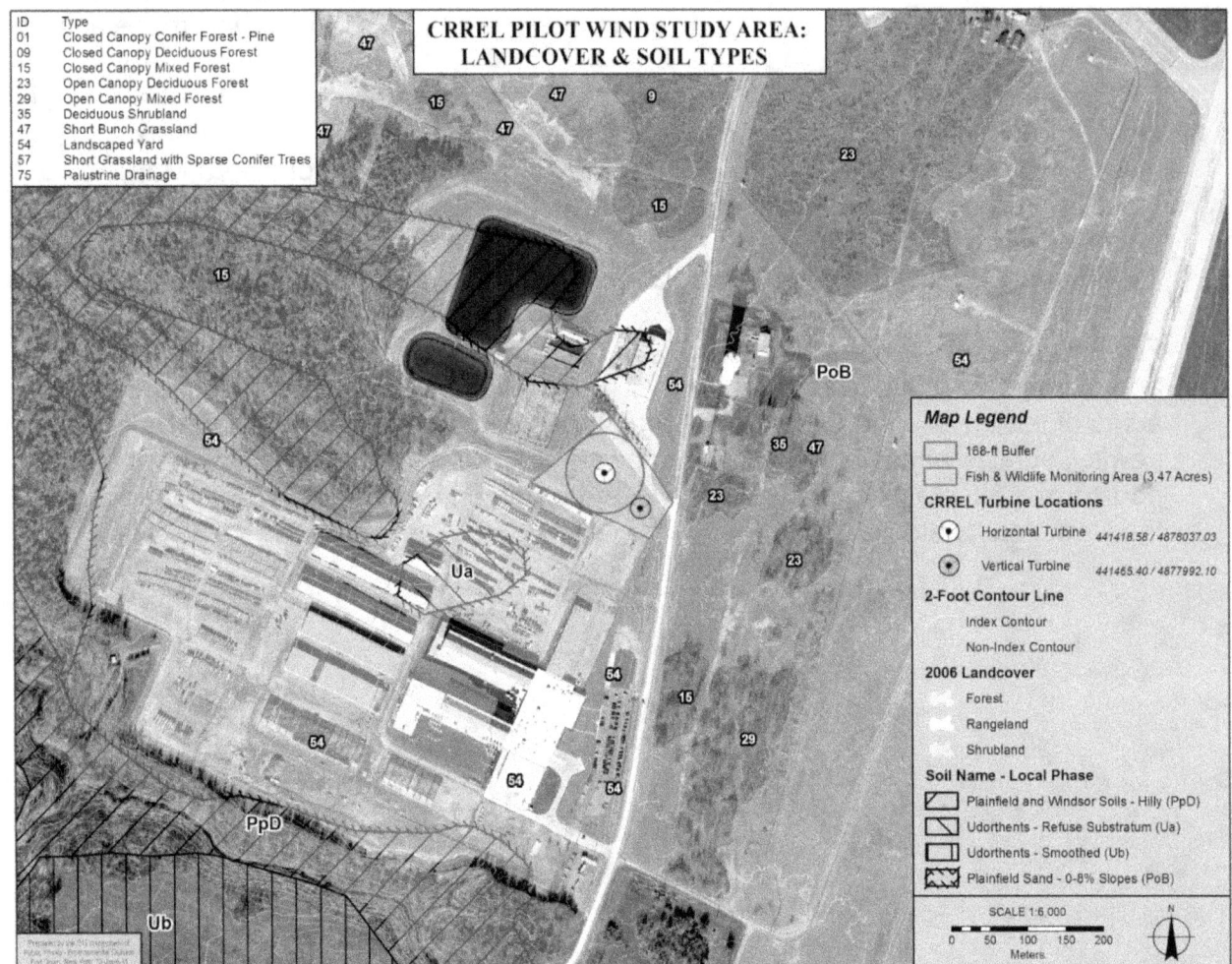

Figure 2.2 - Preferred Study Site

2.3 ALTERNATIVES

This site meets all criteria for the project location where the greatest amount of data could be gathered in close proximity to the radar tower array, easily accessible, near the power grid, in an open space that would accommodate both towers and the area defined by necessary standoff distances for monitoring bat, bird, and ice concerns. The location with the least number of issues is the site located south of the Installation's Central Vehicle Wash Facility (CVWF). It meets the siting criteria

Alternative 2: Division Hill Alternative

The establishment of the study site on Division Hill, though meeting most of the siting criteria, has been rejected because of the proximity to the airfield accidental potential zone and radar issues and will be removed from further analysis as an alternative in this EA.

Alternative 3: Range 2 or Range 1 Alternative

Both Range 1 and Range 2 locations were asked to be considered for this study but are found to be in close proximity to high use areas of the federally endangered Indiana bat (Myotis sodalis). Without a better understanding of potential impacts to bats from these types of turbines, the turbines may have to operate only during limited times to avoid unnecessary impacts to the Indiana bat population. Having restrictions on the amount of time, and or the time of year the turbines can be operated would limit the utility of the study. These sites are being removed from further analysis for this study since under the aforementioned restrictions and because the potential risk for an Indiana bat strike is high neither Range 1 nor Range 2 are acceptable locations.

Alternative 4: Range 48 Alternative

The establishment of the small wind study at Range 48 could be quite viable form an environmental perspective of seeing the affects to birds and bats of small wind with limited potential for a violation of ESA. However the wind strength and direction and the fact that the original study goal was to explore the use of small commercially available wind systems in the Cantonment Area at a large military installation makes this alternative less attractive to the CREL whom is funding the research and is therefore removed as a study location.

Alternative 5: No Action Alternative

Under the No Action Alternative there would be no investigation of the viability of small wind turbines at Fort Drum. Without studying new sources of energy the installation is compromising the Army and DoD directives for sustainability and energy conservation. The No Action Alternative is not a viable means for meeting the current and future requirements for a clean energy economy and energy secure future as required by Executive Order 13541.

This alternative does not meet the purpose and need for the Proposed Action and therefore is not a feasible alternative. This alternative is included as required by Council on Environmental Quality (CEQ) and Army NEPA-implementing regulations. The No Action Alternative provides a benchmark to compare the magnitude of the environmental effects of the Proposed Action / Alternatives

3.0 AFFECTED ENVIRONMENT AND ENVIRONMENTAL CONSEQUENCES

3.1 Setting

Fort Drum is located in northwestern New York State in Jefferson and Lewis counties. About 83 percent of Fort Drum is in the northeastern corner of Jefferson County with the remainder in the northwestern corner of Lewis County. St. Lawrence County borders the installation to the north. The Cantonment Area is about six miles east of Interstate Highway 81 and about 10 miles northeast of the City of Watertown. Fort Drum is served by several state roads and has an extensive local road network. Most of the installation extends northeastward from the Cantonment Area, averaging about 10 miles wide and 20 miles long. Lake Ontario is about 20 miles west of the installation, and the St. Lawrence River is about 20 miles to the north. Fort Drum encompasses 107,265 contiguous acres (167.6 square miles).

3.2 Air Quality and Climate

3.2.1 Fort Drum is under the jurisdiction of the NYSDEC Region 6 office located in Watertown. Compliance with national ambient air quality standards (NAAQS) is determined through the use of ambient air monitoring stations located throughout the state, including monitors in the vicinity of Fort Drum. Fort Drum and Jefferson County is designated as a moderate ozone non-attainment area for 8hr ozone. All other criteria pollutants have been designated as being in attainment.

Based on the Clean Air Act (CAA) Amendments of 1990, a major source is defined as one that emits more than 100 tons per year (ty^{-1}) of criteria pollutants (carbon monoxide, nitrous oxides, sulfur oxides, and PM_{10}), more than 10 t/y of any individual hazardous air pollutant, or more than 25 t/y of any combination of hazardous air pollutants from stationary sources. Actual emissions from stationary sources at Fort Drum fall below the thresholds while the potential-to-emit (PTE) levels that exceed these threshold values exist (U.S. Army 2001). Fort Drum currently operates under a Title V CAA permit. The wind facilities to be constructed will need to be quantified and potentially included in the Title V inventory. The RONA analysis requirements for construction are included in the tables that follow. The equipment list and reference numbers reported are based on construction of a small wind per-turbine number, and are taken from a Programmatic EA for Wind Energy USMC April 2011.

Conformity Analysis		Reference HP of equipment used in past			Emission Factor Based g/hpHr			Emissions T/Yr	
Equipment Type	Number ea	HP	Fuel	Load Factor	VOC	NOx	Estimated HR of use	VOC	NOx
Tractor/Backhoe	2	108	D	55	1.19	7.16	240	0.02	0.1
Crawler/Scrapers	1	479	D	57.5	0.57	5.55	120	0.01	0.06
Watertruck	1	250	D	50	0.57	5.55	120	0.01	0.08
Dump Trucks	1	479	D	57	0.57	5.55	120	0.02	0.17
Trencher	1	63	D	75	1.47	8.72	60	0	0.02
Compactors	1	8	D	43	0.68	4.33	60	0	0
Crane	1	399	D	43	0.63	6.27	120	0.01	0.12
Fork Lift	1	93	D	60	1.28	7.55	120	0.01	0.05
Excavator	1	168	D	57	0.59	6.15	120	0.01	0.07
Drilling Rig	1	291	D	75	0.7	6.71	120	0.02	0.17
Bobcat	1	44	D	55	2.25	5.68	120	0.01	0.02
Compressor	1	106	D	48	1.32	7.76	120	0.01	0.05
Concrete Truck	1	210	D	20	0.57	5.55	120	0	0.03
							Total	0.13	0.77

Emission for Commute	Light Duty Truck
Vehicles	15
Speed	33
Mi/vehicle/day	40
miles/day	10
Emissions Factor lb/day	
VOC	0.2
NOx	0.39
Total emissions T/Yr	
VOC	0.003
NOx	0.0059

Emissions in Tons/Year - Construction		
	VOC	NOx
Emission for Construction	0.13	0.77
Emission for Commute/Lunch	0	0.00585
Total Construction Vehicles	0.16	0.77585

3.2.2 Fort Drum's climate is more fully described in Section 5.6 of the INRMP. Fort Drum has a primarily humid, continental climate with long, cold winters and short, warm, and often humid summers. The mean annual temperature at Fort Drum, averaged over a 10 year period, at Wheeler-Sack Army Airfield (WSAAF), is 48 °F. January is the coldest month, closely followed by February and December. Temperatures fall below 0 °F 24 days/yr on an annual basis; below-freezing temperatures occur on average 148 days annually. With a higher elevation and a greater distance from Lake Ontario, the northeastern part of the installation has average winter temperatures 2-4 °F lower.

Winter temperatures can be severe and with wind chills can fall below -40°F. The warmest months are June, July, and August, with mean monthly maximum temperatures of 74, 79, and 77°F, respectively.

The mean annual precipitation on Fort Drum is about 42 inches, and precipitation is well distributed throughout the year. Snowfall is fairly heavy, with an annual average of 109 inches at Fort Drum. Snow cover can extend from December through March.

Wind velocities on Fort Drum are moderate, averaging seven knots over a 10 year period. The consistent wind is one of the reasons to explore small wind viability on Fort Drum The most violent winds are those that may accompany thunderstorms in late spring, and severe winds of 40-50 knots or more occur once or twice annually on average the determination of turbine type will be influenced by frequency and occurrence of high winds.

3.2.3 Environmental Consequences: Air Quality and Climate

Proposed Action

The Proposed Action would have short-term temporary minor effects on air quality, the net potential cumulative effect would be beneficial due to the decrease in the GHG and other emissions that contribute to climate change from the extraction and utilization of fossil fuels for power generation. Air quality impact activities would need to be quantified and possibly included in the Installation's Title V permit reporting. When equipment requirements and use periods are determined the extent of the increased output can be quantified in accordance with Title five requirements. The Proposed Action would have minor effects during the construction and site set up but in the use of the turbines the effects would be beneficial to air quality overall.

No Action

The No Action alternative would have no consequences or impacts on air quality or climate.

3.3   Geology and Soils

Fort Drum is covered mainly by deltaic and lacustrine clay/silt deposits resulting from glacial and post glacial events. An important hydro-geologic feature in the south-central portion of Fort Drum is a sand plain known as "Pine Plains". This sand plain is a delta of fine sand that was deposited by the Black River into glacial Lake Iroquois during the last Wisconsin glaciation. It forms a large surficial aquifer. In the northern portion of Fort Drum, metamorphic / Precambrian bedrock is overlain by thin lacustrine deposits of clay or silty clay.

Soils of Fort Drum are generally developed from deltaic/lacustrine or glacial deposits. They have been mapped at the soil sub-series level in Jefferson County (USDA 1989) and at the soil association level in Lewis County (Natural Resources Conservation Service 1960). Soils for wind mill locations are shown in Figure 2.2. Soils in the area are generally deep sand and glacial till well drained; soil permeability is high and fertility low.

In general, soils of Fort Drum can be grouped under the Gray Brown Podzolic Soils and the Podzols, with the Vergennes Association, Adams Croghan Association, and Panton-Vergennes Rockland Association being most prominent. Natural fertility of most soils on Fort Drum is low, and organic soils in other than wetland areas are rare. Rhinebeck Series soils (having slopes of 3% or less in areas not considered urban or built up) located on the northwestern one-third of the installation are considered prime farmland (USDA 1989).

Soils at the preferred location include: Plainfield and Windsor soils - Hilly (PpD), Udorthents - Refuse Substratum (Ua), Udorthents - Smoothed (Ub), and Plainfield Sand - 0-8% slopes (PoB). All soils are predominantly well drained and fertile. Soils at this location have historically been disturbed by construction and training activities.

3.3.1 Environmental Consequences: Geology and Soils

Proposed Action

The Proposed Action includes use of the policies and directives set by the Fort Drum Integrated Natural Resources Plan (IMRMP) for the evaluation of land use effects, and maintenance and repair of damaged lands. Establishment of a storm water plan and/or Best Management Practices (BMPs) on site should minimize any erosion or soil impacts and this plan would contain erosion-monitoring requirements for both construction and site BMPs.

No Action

The No Action alternative offers a less comprehensive program for the control and repair of erosion and damage to soils than the Proposed Action at the site location. Consequently, potential minor soil impacts from training and other actions would be greater with the No Action alternative than under the Proposed Action.

3.4   Biological Resources

To ensure sound natural resources management, an Integrated Natural Resources Management Plan (INRMP) was developed and first implemented in 2001. The INRMP was prepared in partnership and signatory cooperation with NYSDEC and the U.S. Fish and Wildlife Service (USFWS), representing the state and federal Sikes Act agencies, respectively. The INRMP and its implementation helps ensure: (1) the sustainability of quality training lands to accomplish the military mission; (2) compliance with environmental laws and regulations; (3) good stewardship of public lands; and (4) enhancement of quality of life on and around Fort Drum. Fort Drum has a staff of natural resources professionals committed to supporting these goals. See the Fort Drum web site to learn more about natural resources management on the installation including fish and wildlife, forestry, and wetlands (http://www.fortdrum.isportsman.net/publications.aspx/, accessed 06/13/2011).

A more detailed description of Fort Drum's biological resources can be seen in sections 5.7 and 5.8 of the INRMP. The biological resources in the study area will change from the current NRMU classification of Short Bunch Grassland to one of Disturbed Developed. The Proposed Action would not have significant negative environmental consequences outside the footprint of the study area.

3.4.1   Flora

Fort Drum has large undeveloped areas with a multitude of biological communities including coniferous and hardwood forests, oak savannahs, shrub-lands, grasslands, and various wetland and open water habitats. There are 993 known plant species that occur on Fort Drum. A current list of Flora found on Fort Drum can be viewed on the interned at http://www.drum.army.mil/garrison/pw/fishandwild.asp. Coastal Environmental Services, Inc. (1993) identified, mapped, and described three exemplary natural communities on Fort Drum. These communities are medium fens, in Training Area 19, northern white cedar swamp, the largest of which is in Training Area 16, and northern sand-plain grasslands in training area 7 and in the vicinity of WSAAF. The Construction of the wind turbines will not affected these communities.

Figure 2.2 Shows the preferred study site and predetermined locations for wind turbines, with the land cover broken out into the NRMU classes found there; The Study site is to be placed on a location classified as disturbed developed. The total area, of the NRMU by type, that will be affected by the proposed action were calculated based on the construction clearing area foot print and the associated fragmentation of the contagious NRMU's and yielded the following totals:, Study site .2A.

3.4.2   Fauna

A multitude of diverse and relatively undeveloped habitat types are found on Fort Drum which, are utilized by a wide variety of species. Numbers of known species occurring on Fort Drum include: approximately 48 mammals, 242 birds, 52 fish, 12 reptiles, and 20 amphibians. The forest, grassland, and wetlands provide habitat for a variety of wildlife animals within this region of the installation.

The Proposed Action and Alternatives would have a minimal impact on the biological resources as a result of the loss of vegetative cover and wildlife habitat for a variety of species. Approximately 90% of Fort Drum is undeveloped and is in close proximity to other agricultural or undeveloped areas including the 5-million acre Adirondack Park, which provides many wildlife species the ability to move freely through large tracts of undisturbed habitat. A complete list of species for the Installation can be found in the INRMP http

Other species are managed in accordance with the installation's INRMP. Compliance with regulations outlined in the BO and INRMP would avoid any significant adverse impacts on these wildlife populations. As a result of these compliance measures and BMPs, the impacts to Fort Drum's biological resources are anticipated to be less than significant.
See the INRMP for all designated state-listed species on Fort Drum.

3.4.3 Sensitive, Threatened, and Endangered Species

Currently there is only one federally-listed species on Fort Drum: the federally endangered Indiana bat (Myotis sodalis). Indiana bats are known to roost and forage on Fort Drum. Indiana bats and other bat species on Fort Drum and across the northeastern U.S. are currently being impacted by White-nose Syndrome. There is no Critical Habitat designated on Fort Drum or anywhere else in New York State for the Indiana bat. An Indiana bat maternity colony is known to exist in the Cantonment Area with the nearest Indiana bat roost located approximately 2700 meters from the preferred location.

Fort Drum completed an informal consultation with the U.S. Fish & Wildlife Service (USFWS) for the construction and operation of these facilities for calendar year 2011. Fort Drum determined that the proposed construction and operation of the wind turbines for calendar year 2011 may affect, but is not likely to adversely affect the Indiana bat, and the USFWS concurred with that determination (Appendix B). Additional analysis of the potential impacts for the operation of the turbines year round (as part of the overall study) will be completed in a new three-year installation-wide biological assessment proposed to start January 2012.

There are 31 known state-listed wildlife species on Fort Drum including 5 endangered, 8 threatened, and 18 species of concern. The five NYS endangered species include: Indiana bat (Myotis sodalis), golden eagle (Aquila chrysaetos), short-eared owl (Asio flammeus), black tern (Chlidonias niger), and peregrine falcon (Falco peregrinus). The eight NYS threatened species include: Henslow's sparrow (Ammodramus henslowii), upland sandpiper (Bartramia longicauda), northern harrier (Circus cyaneus), sedge wren (Cistothayus platensis), bald eagle (Haliaeetus leucocephalus), least bittern (Ixobrychus exilis), pied-billed grebe (Podilymbus podiceps), and Blanding's turtle (Emydoidea blandingii).

Both eagles have been observed migrating over the site but are not known to nest locally. Upland Sandpipers nest nearby in the vicinity of the airport, where migrants also gather during the late summer, but use of the actual site is unlikely because of the presence of woody vegetation and development. (Time of year restrictions for land clearing will minimize direct take of any migratory bird species.)Fort Drum is not required to afford state-listed species any special protection based on their status by the state.

3.4.4 Environmental Consequences: Biological Resources

Proposed Action

The Proposed Action would provide mitigation measures and management of faunal and floral resources around wind study areas. The Fort Drum INRMP uses an ecosystem management strategy to preserve the biological diversity on federal lands, in accordance with the Department of Defense Biodiversity Initiative (The Keystone Center 1996). It emphasizes the use of native species, as emphasized on the Presidential memorandum to the heads of all federal agencies (Office of the President 1994).

The proposed action will change areas of Fort Drum from their current NRMU land classification to one of disturbed developed in the proposed location. The use of the BO reporting requirements and the INRMP result in a formalization of the procedures that will be used to protect the natural systems from deleterious impacts in those areas. The INRMPs policies and procedures, as well as the requirements of, 32 CFR 651 Appendix C Mitigation and Monitoring identify five approaches to mitigation of impacts. They include avoidance and minimization of impacts, rectifying impacts, reduction and compensation for impacts. The procedures established in the NEPA guidelines and the INRMP and the utilization of them for monitoring both effectiveness and enforcement, as required by 32 CFR 651 Appendix C section (e) requirements can and will allow for wind study to proceed while protecting species of concern and maintaining the natural systems in the area to allow for their continued sustainable military use in the future.

The potential direct effects of the construction of the small wind turbines (80 -100 foot tall towers) on bats (including Indiana bats) should be negligible, as there will be no tree clearing required for support of the preferred alternative. Additionally, because tilt type mono pole construction will be utilized with no guy wires, the mortality or wounding events to bats from the establishment of the towers will be limited.

Although there are currently no anticipated effects to bats from the operation of the turbines, they will only be operated during October 1- April 15 or during daylight hours before 1 October until further analysis and evaluation can be completed. Both turbines are equipped with a programmable brake to stop the turbine at or during specific times. There is still a small possibility that migratory bats such as red (Lasiurus borealis) or hoary (Lasiurus cinereus) bats may be impacted during the early fall months as they migrate, however, the potential impacts are anticipated to be negligible.

The potential effects of the construction of small wind turbines (80 -100 foot tall towers) on birds are not known, as there have been no large scale mortality studies performed at small wind facilities, however, researchers studying one tower in Pennsylvania concluded that the prospects of small turbines on short towers killing birds are very low (Andersen 2008). Other researchers on bird mortality at wind facilities have expressed their opinions that short turbines are unlikely to kill many birds because migrating birds typically fly 500-1000 feet above the ground (Bill Evans, Pers. Comm.).

Fort Drum Fish and Wildlife personnel are collecting bird occurrence and abundance data at the Tank Trail site as preconstruction data to evaluate the potential for impacts to migratory birds. Bird data collected at the site will include raptor surveys to document what species migrate over the site and in what numbers, and breeding bird surveys to determine the suite of breeding species at the site.

The fish and wildlife and NEPA program's will be utilized to monitor impacts and effects on the wind turbine during construction of the site and during operation which after the Fort Drum BA is updated and BO received from the USFWS it may allow for the inclusion of full years operation and scientific analysis of the impacts of small wind on birds and bat populations. The information, procedures and plans mentioned in this section will allow Fort Drum to comply with the 32 CFR 651 Appendix C Mitigation and Monitoring requirements of the proposed action.

No Action

The No Action alternative would be less effective than the Proposed Action since it would emphasize reaction to problems rather than a proactive approach. Implementation of the no Action alternative would emphasize responses to current needs to support the military mission as well as site-specific responses to environmental compliance. Reactive management would probably achieve compliance with laws, but it would not provide any long-term benefits to knowledge base of the impacts to biological resources of small wind.

3.5   Wetlands and Water Resources

Wetlands on Fort Drum, including the main impact area, have been identified by six sources, the National Wetland Inventory in 1981 (14,089 acres including open waters), New York State Wetland Survey in 1986 (6,036 acres, did not include any wetlands below 12.5 acres), vegetation/land cover mapping by Coastal Environmental Services in 1992 (12,711 acres), and the U.S. Army in 1996 (15,772 acres) and the Natural Resources Management Units (NRMU) data base project of 2005 which lists 13,742 acres as flooded, saturated or wetland habitats and 19,629 acres with all open water areas included. Wetland types including forested wetlands, freshwater marshes, riparian areas, scrub-shrub wetlands, and wet meadows are found in all areas of the installation.

3.5.1   Environmental Consequences: Wetlands and Water Resources

There are no wetlands or surface water resources on the preferred site location. The nearest wetland/surface water is located .approximately 300 meters from this location.

Proposed Action

The Proposed Action includes the use of the INRMP provisions and storm-water BMPs for planning and evaluating land use effects, and management and repair construction activities. Brief periods of increased sedimentation are likely during construction activities, but these would be minimized by the use of BMPs. The Proposed Action offers the most effective mitigation for damages incurred to surface waters. Implementation of the Proposed Action would not significantly affect groundwater.

The proposed action will not require the preparation of a Clean Water Act (CWA) Storm Water Pollution Prevention Plan (SWPPP) and the incorporation of the project area in the post wide drainage study.

There are no foreseen wetland impacts for this proposed project. A wetland delineation will be performed to identify any wetlands within the study site and any wetland impacts that may occur and require mitigation will be mitigated appropriately as defined under section 404 of the Clean Water Act.

No Action

The No Action alternative or maintaining the status quo offers a less comprehensive program than the Proposed Action for the control and repair of the area. The No Action alternative would only postpone the development of an area slated for use. Consequently, the area will be developed sometime in the future if not as part of this action. No CWA permits are required under the No Action.

3.6   Cultural Resources

Cultural resources include prehistoric and historic archaeological sites, buildings, structures, districts, objects, artifacts, or any other physical evidence of human activity considered important to a culture, subculture, or community for historic, traditional, religious, scientific, or other reasons. For ease of discussion, cultural resources are divided into archaeological resources, historic architectural resources, and traditional cultural properties. Historic properties include cultural resources that are listed in, or eligible for, the National Register of Historic Places.

Cultural resources at Fort Drum are managed according the 2011-2015 Fort Drum Integrated Cultural Resources Management Plan (ICRMP), which is updated every five years. Program features include an up-to-date list of cultural contexts, an archeological site sensitivity map based on a sophisticated predictive model, and detailed tracking of all cultural resource activities through Geographic Information Systems (GIS) and a relational database system. Fort Drum is able to integrate the management of its cultural resources with its mission activities because of this installation-specific cultural resources management program.

As indicated in the ICRMP, Fort Drum has completed archeological inventory of approximately 87 percent of its surveyable acreage, excluding the permanent impact areas and the previously developed portion of the Cantonment Area. The archeological survey completed on approximately 69,000 acres thus far has identified a total of 891 sites that began with earliest human occupation of the region approximately 11,000 years ago and continue through construction of World War II military training features in the 1940s.

Fort Drum currently tracks a total of 940 archeological sites, 1 district with standing structures, and 5 archeological districts, and supports management of 13 historic cemeteries. Resources of concern include the historic districts, 2 traditional cultural properties (TCP), 13 cemeteries and an as yet undetermined number of archeological sites considered eligible for listing on the National Register of Historic Places (NRHP).

3.6.1 Environmental Consequences: Cultural Resources

Proposed Action

Site investigations for the proposed establishment of a new wind turbines resulted in 0 shovel test required. The INRMP and ICRMP include procedures for the protection of cultural resource sites during implementation of projects. Ground-disturbing projects in un-surveyed areas must have site-specific surveys prior to implementation. Review of projects by the Cultural Resources Program Manager and the NEPA process are used to ensure protection of known and potential cultural resources. Mitigation of impacts would be implemented as required in the Fort Drum INRMP as part of the mitigation plan for the project.

No Action

The No Action alternative would have no negative effects on cultural resources since Fort Drum would still have to comply with laws and policies related to cultural resource surveys in un-surveyed areas prior to implementation of projects.

3.7   Noise

Noise is usually defined as unwanted sound, a definition that includes both the psychological and physical nature of the sound (AIHA 1986). Under certain conditions, noise may cause hearing loss, interfere with human activities at home and work, and may affect a person's health and well being in various ways.

Sound pressure level (LP) can vary over an extremely large range of amplitudes. The decibel (dB) is the accepted standard unit for measuring the amplitude of sound because it accounts for large variations in amplitude and reflects the way people perceive changes in sound amplitude. Sound levels are easily measured, but the variability is subjective, and physical response to sound complicates the analysis of its impact on people. People judge the relative magnitude of sound sensation by subjective terms, such as "loudness" or "noisiness".

The human hearing system is not equally sensitive to sound at all frequencies. Because of this variability, a frequency-dependent adjustment called the A-weighting has been devised so that sound may be measured in a manner similar to the way the human hearing system responds. The use of the A-weighted sound level is abbreviated "dBA."

Community noise levels usually change continuously during the day. However, community noise exhibits a daily, weekly, and yearly pattern. Several descriptors have been developed to compare noise levels over different time periods. One descriptor is the equivalent sound level (Leq). The Leq is the equivalent steady-state A-weighted sound level that would contain the same acoustical energy as the time varying A-weighted sound level during the same time interval. Another descriptor, the day-night average sound level (DNL), was developed to evaluate the total daily community noise environment. DNL is the energy average A-weighted acoustical levels for a 24-hour period with a 10 dB upward adjustment added to the nighttime levels (10:00 p.m. to 7:00 a.m.).

The degree of annoyance has been found to correlate well with the DNL based on long-term exposure. Annoyance for short-term activities, such as construction noise, could be influenced by other factors such as awareness and attitude toward the activity creating the noise (U.S. Army 2001b).The construction noise of this action will be short in duration and intensity.

Typical noise sources in and around Fort Drum usually include aircraft, artillery and blast, surface traffic, and other human activities. Major noise contributors at Fort Drum are Army ground weapon firing and the impact of the projectile; Army, Air Force, and Air National Guard fixed-wing and rotary-wing aircraft; and the impact of air-to-ground weapons. Artillery weapons typically generate the highest noise levels; however, the highest sound exposure levels generated by single events are attributed to aircraft over flights.

Noise contours depicting cumulative exposure during an average annual day are the principal analytical tool. The predicted noise DNL for each of the wind turbines is less than 65 dBA, See turbine spec sheets (Appendix E), and the location of the study turbines will not be in a residential area. No significant noise impacts are predicted either during construction or operation of the wind turbine s facilities.

3.7.1 Environmental Consequences: Noise

Proposed Action

The Proposed Action would have negligible impacts on noise. However, the Proposed Action alternative would have minor effects on air quality, the net potential cumulative effect would be beneficial due to the decrease in the GHG and derivative of other emissions that contribute to climate change from the extraction and utilization of fossil fuels for power generation Air quality impact activities would need to be quantified and possibly included in the Title V permit reporting. When equipment requirements and use periods are determined the extent of the increased output can be quantified in accordance with Title five requirements.

No Action

The No Action alternative would have no consequences or impacts on Air Quality or noise.

3.8   Socioeconomics

Fort Drum provided a significant contribution to the economies of Jefferson, Lewis, and St. Lawrence counties in fiscal year 2008 (FY08). Fort Drum's total dollar impact on the surrounding area was estimated at $1,682,987,413. Included in this amount are expenditures for payroll, construction and service contracts, veterinary expenses, estimated contractor payrolls associated with new construction at Fort Drum, direct dental and medical expenses, contribution to local charities, and education expenses (e.g., tuition assistance, contracts). The total payroll (i.e., military and civilian) at Fort Drum was estimated at $1,023,893,471 in FY08 (US Army, 2010).

Approximately 22,805 persons were employed by Fort Drum in FY09. Post employment consisted of 18,023 military employees (81 percent) and 4,782 civilian employees (19 percent) by the end of FY09 (US Army 2011). Due to Army Transformation efforts at Fort Drum, the number of Soldiers has grown by nearly 8,000 between 2003 and 2008 (US Army, 2010).

Fort Drum awarded $45,213,833 in construction contracts (from a total of $72,838,395 awarded) to companies in Jefferson, Lewis, and St. Lawrence counties during FY08. Fort Drum also awarded nearly $14.4 million supply and service contracts to businesses in the tri-county area during FY08 (US Army, 2010).

Due to ongoing deployment activities, the population at Fort Drum varies significantly due to frequent fluctuation of Active Duty and Reserve component units training and deployment cycles.

The U.S. Census Bureau information shows the 2009 population for the tri-county area was 254,591 (Jefferson County - 118,719; Lewis County - 26,157; St. Lawrence County - 109,715). The population of the tri-county area grew by 1.3 percent from 2000 (US Census Bureau, http://factfinder.census.gov/ accessed 03/16/2011).

3.8.1 Environmental Consequences: Socioeconomics

The Proposed Action and the No Action alternative would have some beneficial consequences and impacts on socioeconomics of the Fort Drum region. Both alternatives provide employment for a number of military and civilian personnel and the associated revenue generated for housing and other necessary products required by those personnel and their families. The proposed alternative in addition would generate income for area companies and businesses associated with the construction of the project. Although both alternatives are beneficial, the No Action alternative offers less beneficial than the Proposed Action because the monies generated during the construction of the wind turbines and the maintenance and monitoring of them.

3.9   Environmental Justice

The purpose of Executive Order (EO) 12898, *Federal Actions to Address Environmental Justice in Minority and Low-Income Populations* is to avoid the disproportionate placement of adverse environmental, economic, social, or health impacts from Federal actions and policies on minority and low-income populations. The first step in the process is to identify minority and low-income populations that might be affected by implementation of the proposed action or alternatives (US Army 2000).

The Region of Influence (ROI) for this action is considered to be located solely within the boundaries of properties controlled by Fort Drum.

3.9.1 Environmental Consequences: Environmental Justice

The President directed each federal agency to make it a high priority to identify and assess environmental health risks and safety risks that may disproportionately affect children. The President also directed each federal agency to ensure that its policies, programs, activities, and standards address disproportionate risks to children that result from environmental health or safety risks. There is no evidence to suggest that the Proposed Action or the Alternatives would have a disproportionate environmental health risk or safety risk to children.

3.10 Protection of Children

Executive Order 13045 seeks to protect children from being disproportionately exposed to environmental health or safety risks that may arise as a result of Army policies, programs, activities and standards.

Historically, children have been present at Fort Drum as residents and visitors (e.g., living in family housing, attending schools, using recreational facilities). The Army has and continues to take precautions for their safety by a number of means to include, fencing, limiting access to certain areas, and providing adult supervision (US Army, 2004).

3.10.1 Environmental Consequences: Protection of Children

The President directed each federal agency to make it a high priority to identify and assess environmental health risks and safety risks that may disproportionately affect children. The President also directed each federal agency to ensure that its policies, programs, activities, and standards address disproportionate risks to children that result from environmental health or safety risks. There is no evidence to suggest that the Proposed Action or the Alternatives would have a disproportionate environmental health risk or safety risk to children.

4.0 Cumulative Impacts

Cumulative impacts were analyzed for each management action by adding past, present, and reasonably foreseeable future actions to the No Action (current management) and the Proposed Action (preferred alternative). In determining the consequences of implementing the Proposed Action, it is assumed implementation of the proposed mitigation and management actions will be made in their entirety. The table below provides a summary organized by resource.

Affected Environment	No Action		Proposed Action	
	Environmental Consequences	Cumulative Impacts	Environmental Consequences	Cumulative Impacts
Air Quality/Climate	None	None	Short-term Minor	Beneficial
Geology/Soils	None	None	None	Minor Impact
Wetlands/Water Resources	None	None	Minor Impact	None
Biological Resources- Flora	None	None	Short-term Minor	Minor Impact
Biological Resources- Fauna	None	None	Minor Impact	Minor Impact
Cultural Resources	None	None	None	None
Noise	None	None	None	None
SocioEconimic	None	None	Beneficial	None
Environmental Justice / Protection of Children	None	None	None	None

Neither the Proposed Action nor the No Action alternative would have significant negative environmental consequences compared to existing conditions. The two alternatives differ in the minor habitat change of the action's footprints, the air emissions and bird impacts.

Council on Environmental Quality regulations (40 CFR Parts 1500-1508) define cumulative impacts as "the impact on the environment which results from the incremental impact of the action when added to other past, present, and reasonably foreseeable future actions regardless of what agency (federal or non-federal) or person undertakes such other actions."

Soil erosion and loss of surface vegetation integrity can significantly affect natural systems. The use of time-of-year clearing restrictions, storm water BMP's and other mitigation measures will minimize potential issues.

The Proposed Action has minimal potential for accidental irreversible or irretrievable impact on endangered species by either significant single actions or cumulative actions. Any potential impacts are significantly reduced by implementation of the requirements of the Fort Drum BO and it provisions.

The monitoring of biological resources through provisions of the INRMP and NEPA is a component of the Proposed Action and will provide quantitative data regarding cumulative impacts to biological resources at the wind turbine location. Monitoring programs will be used to adjust management requirements for biological resources, protect ecosystems and assist in managing the lands of Fort Drum to maintain the sustainable use and military training objectives.

The Proposed Action has minimal potential for accidental irreversible or irretrievable commitment of cultural resources by either significant single actions or cumulative actions. This potential will be significantly reduced by careful site selection and implementation of the Integrated Cultural Resources Management Plan (ICRMP).

Cumulative impacts under the No Action alternative would be similar to those of the Proposed Action due to the fact that the proposed location will eventually be developed. However, the Proposed Action would cause the development of this location to occur earlier than the no action alternative. The No Action alternative would continue to negate wind as a potential alternative energy resource. The cumulative impacts on the resources will be minimized through the INRMP approach, monitoring and mitigation of various minor predicted impacts.

4.1   Findings and Conclusions

Fort Drum has implemented an Integrated Natural Resource Management Plan and is updating it for the period 2010-2015. The current and proposed updated plan provide the blue print used to manage natural resources, support the military mission, mitigate environmental effects of the overall military mission and post expansions at the same time complying with various environmental laws the use of the INRMP will ensure that appropriate mitigation and minimization measures are implemented by adopting an integrated, proactive plan to conserve, protect, and improve existing natural resources consistent with existing land uses and the military mission.

Implementing the wind turbine study would result in no significant detrimental impacts to environmental systems as long as stated minimization and mitigation measures are incorporated. Nor would it have significant detrimental impacts on climate, air quality, noise, socioeconomics, or children. Minor adverse impacts on habitats and fauna would be monitored and/or mitigated by full implementation of restorative and proactive environmental management provisions in the INRMP and the USFWS BO requirements. Therefore resulting in beneficial consequences from the project, such as information gained through the analysis of wind turbine radar issues and bird strike impacts of small wind. Impacts to bat will not be analyzed in the current study time frame these issues must be addresses in the reviews Fort Drum BA and the associated USFWS BO. Therefore through built-in mitigation and monitoring requirements the study should avoid violations of federal and state laws, including the Sikes Act, ESA,

CWA, MBTA, CAA and NEPA.

4.1.1 Mitigation Summary

VECs:	No-Action Alternative: Built-In Mitigation:	Proposed Mitigation:	Preferred Alternative: Built-In Mitigation:	Proposed Mitigation:
Air Quality (Construction, Operation, Closure):	N/A	N/A		Paved access area, seeding and mulching disturbed area to stabilize soils
Wetlands and Water Resources:	N/A	N/A	SWPP, Avoidance, BMPs to Prevent Runoff	N/A
Soil Resources:	Soil and erosion control BMPs, Re-seeding	N/A	Soil and erosion control BMPs, Re-seeding	N/A
Biological Resources:	N/A	N/A	INRMP, Avoidance, and Compliance with Fort Drum BO and the ESA, Compliance with MBTA and TOY restrictions	Monitoring of bird strikes operational time frame limited to avoid bat foraging times
Cultural Resources:	Compliance with Section 106 of NHPA, Consultation with NYSHPO and tribes	Avoidance	Compliance with Section 106 of NHPA, Consultation with NYSHPO and tribes	Avoidance
Land use:	N/A	N/A	N/A	N/A
Noise:	N/A	N/A	Offset distances	Locations will have offsets
Socioeconomics / Environmental Justice:	N/A	N/A	N/A	N/A
Traffic / Transportation:	N/A	N/A	N/A	N/A

Implementation of the Proposed Action if mitigation measures are followed would not constitute a Federal action that has significant affect on the human environment. A

Finding of No Significant Impact should be published.

5.0   LIST OF PREPARERS

Walker R Heap III, General Biologist, NEPA, Fort Drum Environmental Division, Adjunct
  Faculty, SUNY JCC; MPS, 2002, Environmental Resource Engineering, SUNY ESF
  at Syracuse University, Syracuse, New York. BS, 1997, Environmental Science and
  Biology, SUNY ESC, AS SUNY Jefferson; Years of Experience: 15

Cait Schadock, NEPA Program Manager, Fort Drum, Environmental Division; BA, 1983,
  Anthropology/Biology, State University of New York, Potsdam, New York; Years of
  Experience: 25.

6.0   REFERENCES

AIHA, 1986. *Noise and Hearing Conservation Manual*, American Industrial Hygiene
  Institute, fourth edition, 1986. AP-42 Volume 2 Appendix H

Andersen, K.W. 2008 (rev. 2009). A Study of the Potential Effects of a Small Wind
  Turbine on Bird and Bat Mortality at Tom Ridge Environmental Center
  Erie, Pennsylvania: A Report Prepared for Pennsylvania Department of
  Conservation and Natural Resources. 16 pp.

Office of the President, 1994. *Environmentally and Economically Beneficial Practices
  on Federal Landscaped Grounds*. Memorandum to Heads of Executive
  Departments and Agencies, April 26, 1994. The White House, Washington, DC.
  3 pp.

Reschke, C., 1990. *Ecological Communities of New York State*. New York Natural
  Heritage Program. New York State Department of Environmental Conservation,
  Latham, NY. 96 pp.

Natural Resources Conservation Service, 1960. *Soil Survey of Lewis County, New
  York*. U.S. Department of Agriculture, Washington, D.C. 107pp.

SUNY Plattsburgh, 1998. *Economic and Demographic Characteristics of Jefferson
County, New York*.

US Army, 1993. *Endangered and Threatened Species Survey on Fort Drum, New
  York*. Prepared by Coastal Environmental Services, Inc. for the Environmental
  Division, Public Works, Fort Drum, NY.

US Army, 1994a. *Cultural Resources of Fort Drum.* Volumes 1-14, Prepared by The Cultural Resources Group, Louis Berger and Associates under contract with the National Park Service, Mid-Atlantic Region.

US Army, 1995. *Environmental Baseline Conditions of Fort Drum, NY.* Prepared for Public Works, Environmental Division, Fort Drum, NY by Parsons Engineering Science, Inc., Liverpool, NY; Fairfax, VA; St. Louis, MO.

US Army, 1996. *Wetlands Mapping Report for United States Army, Fort Drum.* Fort Drum, New York. 33 pp.

US Army, 2001. *Integrated Natural Resources Management Plan 2001-2005, Fort Drum, New York.* Prepared by Universe Technologies, Incorporated, Frederick, MD and Gene Stout and Associates, Loveland, CO for the Environmental Division, Natural Resources Branch, Public Works, Fort Drum, NY. 158 pp. + appendices.

US Army, 2001a. *Finding of No Significant Impact for the Fort Drum Water and Wastewater Utility Privatization Project.* Prepared for Public Works, Environmental Division, Fort Drum, NY by Parsons Engineering Science, Inc., Liverpool, NY; Fairfax, VA; St. Louis, MO.

US Army, 2000. Programmatic Environmental Assessment for Fort Drum, New York. Parsons Engineering Science, Inc., Liverpool, New York. October 2000

US Army Engineer Topographic Laboratories, 1977. *Terrain Analysis, Fort Drum, New York.* The Terrain Analysis Center, Fort Belvoir, VA. 45 pp.

US Army Environmental Center, 1997. *Guidelines to Prepare Integrated Natural Resources Management Plans for Army Installations and Activities.* Aberdeen Proving Ground, MD. 37 pp.

US Army, 2004. Key Transformation- Related Lessons Learned Concerning the Tactical Unmanned Aerial Vehicle for the 172nd Styrker Brigade Combat Team. Center for Army Lessons Learned (CALL), Fort Leavenworth, Kansas.

US Army, 2005 Environmental Assessment for Army Transformation Implementation at Fort Drum, New York. Parsons Engineering Science, Inc., Liverpool, New York. October 2005.

US Army, 2010. *Fort Drum Economic Impact Statement, Fiscal Year 2009, October 1, 2008 - September 30, 2009.* United States Department of the Army. April 2010.

US Army, 2011. *Integrated Natural Resources Management Plan 2011-2015, Fort Drum, New York.* Prepared by Directorate of Public Works Environmental Division Natural Resources Branch, Fort Drum, NY. April 2011.

US Army, 2011. *Environmental Assessment for Stationing Actions to Support the Grow the Army Initiative at Fort Drum, NY.* Prepared by the Environmental Planning Branch Environmental Quality Programs Division US Army Environmental Command, San Antonio TX, for the Directorate of Public Works Environmental Division Natural Resources Branch, Fort Drum, NY. February 2011.

USDA, 1989. *Soil Survey of Jefferson County, New York.* U.S. Department of Agriculture, Washington, D.C. 376pp

APPENDIX A

REVIEWERS AND AGENCIES CONTACTED

Reviewers and Agencies Contacted

James Miller	- Chief, Environmental Division Fort Drum
Jason Wagner	- Chief Natural Resources Branch, Fort Drum
Cait Schadock	- NEPA Coordinator, Fort Drum
Walker Heap	-NEPA Biologist, Fort Drum
Steve Rowley	-Energy Program Manager
Joe Donnelly	- Range Operations, Fort Drum
Chris Dobony	- ESA Biologist, Fort Drum
Rich Falcon	- RTLA Coordinator, Fort Drum
Donna Mansulla	- Hazardous Waste Program Manager Fort Drum
Jason Murray	- Wetlands Regulatory Program, Fort Drum
Franklin Page	- Air Program Manager Fort Drum
Raymond Rainbolt	- Fish and Wildlife Program Manager, Fort Drum
Anthony Rambone	- POL Program Manager Fort Drum
Laurie Rush	- Cultural Resources Program Manager, Fort Drum
Scott Siegfried	- Wetlands Program Manager, Fort Drum
Rodger Voss	- Installation Forester, Fort Drum
Ian Warden	- LRAM Coordinator, Fort Drum
Paul Zang	- Chief Compliance Branch, Fort Drum

APPENDIX B

CORRESPONDENCE

Concurrence letter from US Fish and Wildlife Service, 27 May 2011.

BO, Letter US. Fish & Wildlife Service, June, 3 2009.

BO, Letter US. Fish & Wildlife Service, March, 24 2009.

Letter of No Effect to Cultural Resources, 12 April 2011

DEPARTMENT OF THE ARMY
HEADQUARTERS, UNITED STATES ARMY GARRISON
FORT DRUM, NEW YORK 13602-5000

To: USFWS-New York Field Office
ATTN: Sandra Doran/Robyn Niver
FAX #: 607-753-9699

From: Fort Drum Fish & Wildlife Mgmt Program
ATTN: Ray Rainbolt / Chris Dobony
FAX #: 315-772-5974

Request for USFWS concurrence with the determination below pursuant to Section 7(a)(2) of the Endangered Species Act of 1973 regarding the Indiana bat (*Myotis sodalis*)

Date: 25 May 2011
Project/Activity: Fort Drum Small Wind Study in TA 4A.

The overall goal of this study is to evaluate small (100 foot towers or under) vertical and horizontal axis turbines for use on Fort Drum. This consultation will assess only the construction and operation of these turbines for Calendar Year 2011. Additional evaluation will be performed during the development of the 2012-2014 Biological Assessment.

The 2011 project will consist of erecting one vertical axis and one horizontal axis wind turbine in Fort Drum's TA 4A (Horizontal-44148.58/4878037.03, Vertical-441465.4/4877992.1). Approximately 2 5 acres of sparse grassland will be cleared to support this activity. An additional fact sheet on the specifications of the two turbines will also be provided to the USFWS as part of this consultation package.

The U.S. Army Fort Drum Garrison has determined the proposed project:

  X may affect, but is not likely to adversely affect   ___ may affect ___ will result in no effect

Justification:

- The proposed construction of the two turbines will not result in any direct effects to Indiana bats. There will be no tree clearing required.
- Tilt type mono pole construction with no guy wires will be used, therefore mortality or wounding events to bats from the establishment of the towers will be limited.
- Although the project site is within known Indiana bat range, there are currently no known Indiana bat roosts nearby. The closest roost is ~2700 meters away.
- Although the project site is within known Indiana bat range and nearby to known foraging areas, there are no known movements of Indiana bats in and around the project area.
- Although there are currently no anticipated effects to bats from the operation of the turbines, they will only be operated during October 1-April 15 or only during daylight hours before 1 October until further analysis and evaluation is completed during the development of the 2012-2014 Biological Assessment. Both turbines are equipped with a programmable brake to stop the turbine at or during specific times.

If possible, we request USFWS concurrence with this determination by COB 27 May 2011.

The U.S. Fish and Wildlife Service:

[X] Concurs with your determination and has no further ESA comments
USFWS Contact(s): Robyn Niver_____   Date: 5.26.11
                            DAS 5/26/11

United States Department of the Interior

FISH AND WILDLIFE SERVICE
3817 Luker Road
Cortland, NY 13045

June 3, 2009

Colonel Kenneth H. Riddle
Armor, Garrison Commander
Department of the Army
U.S. Army Installation Management Command
Headquarters, United States Army Garrison, Fort Drum
10000 10th Mountain Division Drive
Fort Drum, NY 13602

Dear Colonel Riddle:

This is in regards to the activities conducted at the Fort Drum Military Installation (Fort Drum) located in the Towns of Antwerp, Champion, LeRay, Philadelphia, and Wilna, Jefferson County, and the Town of Diane, Lewis County, New York, and their effects on the Federally-listed endangered Indiana bat (*Myotis sodalis*). As you are aware, in accordance with Section 7 of the Endangered Species Act (ESA) of 1973, as amended (16 U.S.C. 1531 *et seq.*), the U.S. Fish and Wildlife Service (Service) completed consultation with the U.S. Army Garrison Fort Drum (Army) for activities proposed on Fort Drum (2009-2011) and issued a biological opinion on March 24, 2009. On April 9, 2009, we received a request from your staff to revise one of the terms and conditions (#17). We agreed with the recommendation and have enclosed a revised opinion to reflect this. In addition, we found some minor typographical errors which we fixed in this version.

Should you have any questions, please contact Ms. Robyn Niver of this office at (607) 753-9334.

Sincerely,

David A. Stilwell
Field Supervisor

Enclosure

cc: NYSDEC, Watertown, NY (A. Ross)
NYSDEC, Albany, NY (P. Nye/A. Hicks)
Army, Fort Drum, NY (J. Corriveau)
COE, New York, NY (J. Connell)
FWS, Hadley, MA (G. Smith)

United States Department of the Interior

FISH AND WILDLIFE SERVICE
3817 Luker Road
Cortland, NY 13045

March 24, 2009

Colonel Kenneth H. Riddle
Armor, Garrison Commander
Department of the Army
U.S. Army Installation Management Command
Headquarters, United States Army Garrison, Fort Drum
10000 10th Mountain Division Drive
Fort Drum, NY 13602

Dear Colonel Riddle:

This is in response to letters dated December 1, 2008, and January 26, 2009, from Mr. James W. Corriveau regarding activities at the Fort Drum Military Installation (Fort Drum) located in the Towns of Antwerp, Champion, LeRay, Philadelphia, and Wilna, Jefferson County, and the Town of Diane, Lewis County, New York, and their effects on the Federally-listed endangered Indiana bat (*Myotis sodalis*). Since the discovery of Indiana bats at the installation in 2006, the U.S. Army Garrison Fort Drum (Army) has worked closely with the U.S. Fish and Wildlife Service (Service) to conserve this species. We congratulate the Army for using a programmatic approach to assess the potential for impacts to Indiana bats from activities on Fort Drum and have been pleased to work with your staff over the past 1½ years on this type of review.

In accordance with Section 7 of the Endangered Species Act (ESA) of 1973, as amended (16 U.S.C. 1531 *et seq.*), enclosed is the Service's biological opinion produced in response to your January 2009 Biological Assessment for activities on Fort Drum between 2009-2011.

After reviewing the current status of the Indiana bat, the environmental baseline for the action area, the effects of the proposed activities on Fort Drum (2009-2011), and the cumulative effects, it is the Service's biological opinion that the action, as proposed, is not likely to jeopardize the continued existence of the Indiana bat. Critical habitat for the Indiana bat has been designated at a number of locations throughout its range; however, this action does not affect any of those designated critical habitat areas and no destruction or adverse modification of that critical habitat is expected.

As mentioned above, the Army has demonstrated a commitment to Indiana bat conservation and we look forward to continued cooperation to protect this species. Should you have any questions, please contact Ms. Robyn Niver of this office at (607) 753-9334.

Sincerely,

David A. Stilwell
Field Supervisor

Enclosure

cc: Army, Fort Drum, NY (J. Corriveau)
NYSDEC, Watertown, NY (A. Ross)
NYSDEC, Albany, NY (P. Nye/A. Hicks)
COE, New York, NY (J. Connell)
FWS, Hadley, MA (G. Smith)

DEPARTMENT OF THE ARMY
US ARMY INSTALLATION MANAGEMENT COMMAND
HEADQUARTERS, UNITED STATES ARMY GARRISON, FORT DRUM
10000 10TH MOUNTAIN DIVISION DRIVE
FORT DRUM, NEW YORK 13602-5046

REPLY TO
ATTENTION OF

IMNE-DRM-PWE   12 April 2011

MEMORANDUM FOR RECORD

SUBJECT: Wind Study EA

1. The Fort Drum Cultural Resources program has evaluated the proposed location of the wind turbines in Training Area 4A.

2. Survey of the proposed locations was completed in 2001 as part of a general survey of the training area. No cultural resources were identified at the proposed wind turbine locations.

3. Consultation was completed in February of 2003 as part of the 2001 annual report. The SHPO chose not to comment.

4. Fort Drum maintains continuous consultation with the three Native American consultation partners. They had no comments on these recommendations.

5. POC for this action is Duane Quates, Federal Archaeologist, (315) 774-3848.

E.W. Duane Quates
Federal Archeologist

APPENDIX C

GENERAL CONFORMITY RULE
RECORD OF NON APPLICABILITY

GENERAL CONFORMITY RULE
(Code of Federal Regulations (CFR), Title 40 Part 51)

The CRREL and the 10th Mountain Division request that the installation support the creation in the Cantonment on Fort Drum of a small wind study and the placement of two wind turbines at the wash rack location on Fort Drum.

Conformity under the Clean Air Act, Section176, has been evaluated for the proposed action in accordance with 40 CFR Part 51. Fort Drum resides in a tri-county region known as the North Country, New York and includes Jefferson, Lewis, and St. Lawrence Counties. The Fort Drum area is classified by the USEPA as a moderate ozone non-attainment area for 8 hr ozone. Estimated direct and indirect emissions for the year with the greatest intensity of work are described in Table 1 and compared to the *de minimis* levels for areas other than serious, severe, or extreme nonattainment areas inside and ozone transport region. All emissions would fall well below the *de minimis* threshold established in 40 CFT 51.853(b) of 50 tons per year for volatile organic compounds (VOCs) and 100 tons per year for nitrogen oxides (NOx). Additionally, the project/action is not considered regionally significant since the pollutant emissions are less than 10 percent of annual county emissions. Therefore, a formal General Conformity Determination is not required for this proposed action.

Table 1

Comparison of the *de minimis* Levels for Criteria Pollutants for Marginal NAAs, and Estimated Emissions of the Proposed Project

Criteria Pollutant	*De Minimis* Level Tons/Year	Estimated Emissions Tons/year
VOC	50	0.016
NO$_x$	100	.77585
CO	N/A	N/A
PM-10	N/A	N/A
SO$_2$	N/A	N/A

Proponent: Receiving Installation:

Steve Rowley

Steve Rowley, PE
Energy Manager - PW
Small wind study
Date: 6/7/11

James M. Miller

JAMES M. MILLER
Environmental Coordinator
Fort Drum, New York
Date: JUN 0 1 2011

APPENDIX D

WIND TURBINE SPECIFICATIONS

Wind Turbine Specifications - Vertical Axis

CleanField Energy
Head Office
1404 Cormorant Road, Unit #6
Ancaster, Ontario
L9G 4V5 CANADA
R&D Office
774 Gordon Baker Road
North York, Ontario
M2H 3B4 CANADA

Turbine Model Number and/or Name **Power Rating @ 11 m/s**	• V3.5 • 2.9 kW nominally • Rated Wind Speed: 12.5 m/s (28 mph) with 3kW rated power (data provided by Cleanfield representative)
Tower Information	• Company can fabricate any type of tower desired for application (see CleanField folder under file: "VAWT - Army Corp letter") • Pictures of towers shown in file: "VAWT - Army Corp letter" • Any tower height available. Custom build the pole mount per installation. • Most of the turbines are on monopole mounts at 30-50 ft • Guy cables not required and not common for 30-50 ft towers
Rotor Diameter & Info	• Rotor Diameter: 2.75 m • Rotor height: 3m
Blade Speed **Usable Speed Range**	• Calculated max blade speed: 54.7 MPH • Survival Wind Speed: 45 m/s (100 MPH) • Turbine will shut down and lock for anything over 170 RPM • "Due to the inertia of the turbine even if the protection acts before the stated RPM limit the turbine can reach 190RPM. There are situations like

	this in very high wind conditions."—as quoted by Cleanfield representative
Bird and Bat Kill Documentation	• "Birds and bats are not affected by this type of turbine. Vertical Axis " – as quoted by CleanField representative • No official documentation provided by company, but see general information regarding birds and bats in folder: "Bird&BatInfo"
Noise Information	• See file: "VAWT- WindTurbineNoiseReport-McLaren-McMaster University.pdf" in Cleanfield folder (page 12)
Vibration Information	• See folder for Cleanfield turbine file: "V3.5VAWTStructualData-Rooftop& Monopole-0309.pdf"
Radar Impact Measurements	• "Unlike the huge wind turbines that can affect radar, these are much smaller units and have virtually no effect." – as quoted by CleanField representative
Blade Material & Information Deicing/Anti-icing	• 3 blades of Reinforced Fiberglass • Currently no issues with icing, see website: http://www.weican.ca/ for additional information • Rain shields around the generator and the generator kick- start can loosen any ice that may form • The great majority of Cleanfield installations are in Northern Ontario and there has been no icing issues
Certifications (Small Wind Certification Council, National Renewable Energy Laboratory, etc)	• 16-Sept-'08. Cleanfield Energy Announces Installation of Vertical Axis Wind Turbines for Small Wind Turbine Certification With Wind Energy Institute of Canada. See website for additional information http://www.weican.ca/projects/ • Certified to: CSA 22.2 No.107.1-01 • Designed to: IEC 61400-2, UL 1741, IEEE 1547

Brake System	• The 3 phase system is UL and CSA certified • Built-in fail safe brake • Electro- Mechanical system
Cost	• Installed price is nominally $28,000
Delivery Time/Lead Time	• Could be available in approx 10 days • "We have them in stock ready to ship. You can have this one pictured (see folder for photo). It comes in three boxes. " – as quoted by CleanField representative
Life	• 25 years with "very little maintenance"
Customer Contacts	• Waiting for list from company
Dealer/Manufacturer locations	• Currently setting up USA manufacturing in South Carolina and Florida • The turbine is made in Canada at this time
Locations of Existing Turbines	• Virginia Tech University, Blacksburg, Virginia, US • HIT Hamilton Incubator of Technology, Hamilton, Ontario, Canada • Sligo, Ireland • Several in Ontario
Power Information	• Max power: 4.5kW
Warranty	• 5 year warranty
Installation Information	• Crane necessary • Would send people from Canada to install the turbine
Additional Information	• Extensive tests conducted in wind tunnel • Rated RPM: 160; Max RPM: 190 • Overall turbine height: 3.11m • Turbine weight: 540lbs

Wind Turbine Specifications - Horizontal Axis

Aerostar

Aerostar, Inc.
PO Box 52
Westport Point, MA 02791
508.636.3192 aerostarwind.com
sales@aerostarwind.com

Turbine Model Number and/or Name	• Aerostar 6m
Power Rating @ 11 m/s	• 7.75 kW @ 11 m/s • Rated annual energy - 12,688 kWh at 5 m/s
Tower Information	• Monopole, guyed, and lattice towers available • Custom heights available (all with tilt option) • Guyed lattice tower options: 80', 100', 120' (all with tilt option) • Free standing lattice tower options: 80', 100', 120', 140' • Tilting monopole with gin pole/winch: 80', 100' and custom options available for this design as well • "Tilt Down" guyed towers must have 4 sets of guy wires arranged 90° apart • Tower is hinged at the base and tilts by means of a removable gin pole using an optional electric winch or using tractor, backhoe, etc.
Rotor Diameter & Info	• Rotor diameter: 22ft or 6.7 m • Swept Area: 380 square feet or 35.3 sq. meters
Blade Speed **Usable Speed Range**	• 125 mph (measured) • Cut-in Speed: 3.6 m/s or 8 mph • Cut-out Speed: 22.5 m/s • Shut-down wind speed: 45 second gust over 50 MPH • Turbine equipped with over speed control system consisting of a fail-safe brake with centrifugally operated tip brakes • Shutdown speed can be adjusted through software in the microprocessor controlled colored touch screen display panel

Bird and Bat Kill Documentation	• "None specifically in regard to Aerostar, but there are publications that discuss how "small wind" turbines do not impact wildlife."—as quoted by Aerostar representative • Aerostar provides some general information about birds and bats on their website, see link: "Aerostar Bird Freindly.mht" in Aerostar folder
Noise Information	• 43 dBA at 100' at 15 MPH
Vibration Information	• See vibration/harmonics video at http://www.aerostarwind.com/index.html# • No other vibration documentation provided by Aerostar
Radar Impact Measurements	• Blades are fiberglass and small diameter with low rpm and have a negligible effect on radar systems.
Blade Material & Information	• 2 Blade teetering rotor, tapered twisted design • Tapered, Twisted Fiberglass
Deicing/Anti-icing	• "Aerodynamic blade design resists ice build-up." – as quoted by Aerostar representative
Certifications (Small Wind Certification Council, National Renewable Energy Laboratory, etc)	• North American Energy Laboratories, Massachusetts • NYSERDA (on approved list for funding) • California Energy Commission • Filing for SWCC in 2011 • Presently no UL approval on induction generator (North American Energy Laboratories Certification)
Brake System	• Failsafe brake and aerodynamic tip brakes • Stop switch for brake • Can program stopping, if desired
Cost	• Turbine package$63,000.00 • Installation cost: $20-$30,000 • Stark Foundation option (see Aerostar folder File: "StarkFoundationsTechnicalBrochure.pdf") approximately $60,000 with standard turbine package
Delivery Time/Lead Time	• Standard lead time is 12 weeks • Special programs may be available for this application/project
Life	• 25 years

Warranty	• Standard 2 year
	• 5 year option available
Dealer/Manufacturer locations	• Currently manufactured in Westport, MA.
	• In process of expansion and acquiring larger local facility.
Customer Contacts	• Still waiting on information from Aerostar
Locations of Existing Turbines	• Suffolk University in Edmunds, Maine (lattice tower)
	• Westport Small wind farm - Noquochoke Orchards
	• Several others listed on website (see Aerostar folder for additional info)
Additional Information	• Complete data monitoring built in to system
	• Built in anemometer
	• Web interface available (built in for 10kW)
	• Induction speed machine; constant speed motor
	• 10kW has a gearbox
	• Self starter by wind; restart wind speed is adjustable
	• Turbine weight: 650lbs
	• For this application/project, there is an option to deal directly with Aerostar instead of going through a dealer
Technical Contact	• Rob Rollins
	• rob@earostarwing.com
Owner/Founder Contact	• Paul Gay
	• 774.201.9100

Estimated Annual Energy Production at Sea Level		
MPH	Month	Year
8	339	4,072
9	527	6,325
10	755	9,055
11	1,010	12,177
12	1,278	15,336
13	1,564	18,766
14	1,854	22,247
15	2,142	25,699
16	2,414	28,972
m/s	Month	Year
4	518	6,213
5	1,057	12,688
6	1,686	20,234
7	2,323	27,872
8	2,899	34,787

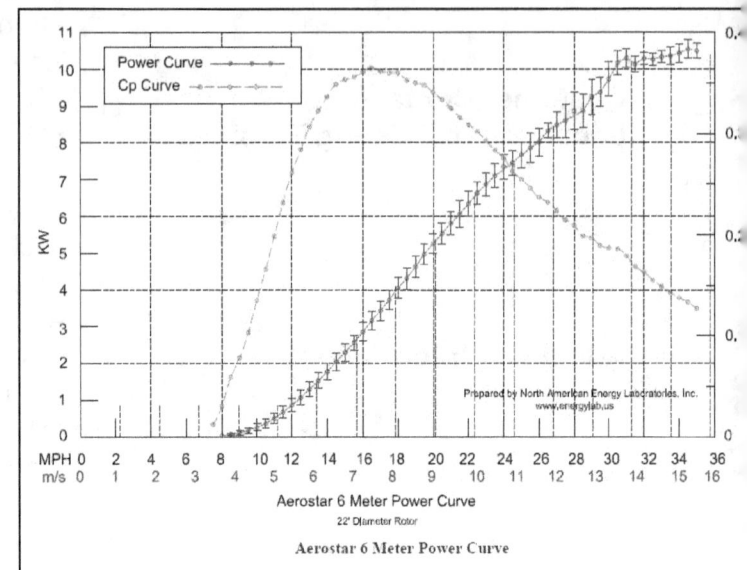

Aerostar 6 Meter Power Curve
22' Diameter Rotor

Aerostar 6 Meter Power Curve

APPENDIX E

DRAFT FINDING OF NO SIGNIFICANT IMPACT

**FINDING OF NO SIGNIFICANT IMPACT
FOR
CONDUCTING A STUDY OF SMALL WIND TURBINES ON
FORT DRUM, NEW YORK**

This Environmental Assessment (EA) documents the results of a study of the potential impacts to the natural and human environment from the construction of and study of small wind turbines on Fort Drum, New York.

This study was conducted pursuant to the requirements of the National Environmental Policy Act (NEPA) of 1969 [42 United States Code, 4321 et seq.], the Council on Environmental Quality (CEQ) Regulations [40 Code of Federal Regulations (CFR) 1500-1508], and 32 CFR Part 651 (a.k.a. Army Regulation (AR) 200-2), Environmental Analysis of Army Actions, Final Rule, 29 March 2003. The purpose of this study was to determine the extent of potential environmental impacts from the proposed action and to decide whether or not those impacts are significant, thereby warranting a more detailed study of possible impacts, mitigation, and alternative courses of action.

The analysis process involved the review of installation natural resources-related data collected by Fort Drum, a variety of other governmental agencies, and private organizations. The process involved interviews with Fort Drum personnel involved with natural resources management, military training & planning, cultural resource management, and operations & maintenance.

This Environmental Assessment (EA) originally included five alternatives three were rejected base on site related preferences and impacts and proximity to the bat conservation area (BCA). A careful review of the elements of the two alternatives performed and assesses the potential impacts of these alternatives. The analysis of impacts (or consequences) of the Proposed Action was based on information about the affected environment on and around the Fort Drum Army Installation as well as on the multiple years of experience of the people involved in the preparation and review of this EA. Following this assessment effort, it is concluded that implementation of the Proposed Action would not have a significant impact on the natural or human environment, as long as measures summarized in the conclusions section of the EA are implemented properly.

Fort Drum proposes construction of and study of small wind turbines. Five Alternatives were considered in this EA including the "No Action Alternative" which serves as a benchmark against which the other alternatives were evaluated. These alternatives are:

Alternative 1: Training Area 4 / CVWF (Preferred Alternative)
Alternative 2: Division Hill Alternative
Alternative 3: Range 2 or Range 1 Alternative

Alternative 4: Range 48 Alternative
Alternative 5: No Action Alternative
Alternatives 2, 3, and 4 were removed from consideration

The EA conclusions, which are incorporated into this Finding of No Significant Impact, examine potential effects of the alternatives on resources and areas of environmental concern that could be affected by this action. These include climate, air quality, and noise, geology and soils, water resources, biological resources, cultural resources, socioeconomics, environmental justice and protection of children.

The analysis determined that implementing the Proposed Action, while modifying the current habitat of the area, this would not have a significant effect on the resource and/or environment. All alternatives provide employment for a number of civilian and contract personnel and the associated revenue generated for housing and other necessary products required by those personnel and their families.

After careful review of the potential impacts of the alternatives, it is concluded that implementing the Proposed Action would not have a significant impact on the quality of the human or natural environment as long as measures summarized in the EA are implemented properly. The Proposed Action has minimal potential for irreversible or irretrievable commitment of natural resources by either actions and or cumulative effects. The action as stated has necessary mitigation and minimization requirements. Because there would be no significant environmental impacts resulting from implementation of the Proposed Action, an Environmental Impact Statement is not required and will not be prepared. This analysis fulfills the requirements of the National Environmental Policy Act and associated Council On Environmental Quality regulations as well as requirements of 32 CFR Part 651 (AR 200-2) Environmental Analysis of Army Actions.

A Public Notice was published in the Watertown Daily Times newspaper 03 July through 01 August 2011 to announce a 30-day public comment period. Copies of the Environmental Assessment and the Finding of No Significant Impact were made available for review upon request. XX comments were received.

_____   Date
Noel T. Nicolle
Colonel, US Army
Garrison Commander

Appendix 2: Federal Aviation HAWT and VAWT Permits

Mail Processing Center
Federal Aviation Administration
Southwest Regional Office
Obstruction Evaluation Group
2601 Meacham Boulevard
Fort Worth, TX 76137

Aeronautical Study No.
2011-WTE-6399-OE

Issued Date: 07/14/2011

Joe White
Joe White
Bldg P-2065 Room 155 Hangar Access Drive
Fort Drum, NY 13602

** DETERMINATION OF NO HAZARD TO AIR NAVIGATION **

The Federal Aviation Administration has conducted an aeronautical study under the provisions of 49 U.S.C., Section 44718 and if applicable Title 14 of the Code of Federal Regulations, part 77, concerning:

Structure:	Wind Turbine Small Wind Turbine Study
Location:	Great Bend, NY
Latitude:	44-03-11.22N NAD 83
Longitude:	75-43-53.40W
Heights:	112 feet above ground level (AGL)
	787 feet above mean sea level (AMSL)

This aeronautical study revealed that the structure does not exceed obstruction standards and would not be a hazard to air navigation provided the following condition(s), if any, is(are) met:

As a condition to this Determination, the structure is marked/lighted in accordance with FAA Advisory circular 70/7460-1 K Change 2, Obstruction Marking and Lighting, red lights - Chapters 4,5(Red),&12.

It is required that FAA Form 7460-2, Notice of Actual Construction or Alteration, be completed and returned to this office any time the project is abandoned or:

_____ At least 10 days prior to start of construction (7460-2, Part I)
__X__ Within 5 days after the construction reaches its greatest height (7460-2, Part II)

While the structure does not constitute a hazard to air navigation, it would be located within or near a military training area and/or route.

This determination expires on 01/14/2013 unless:

(a)   extended, revised or terminated by the issuing office.
(b)   the construction is subject to the licensing authority of the Federal Communications Commission (FCC) and an application for a construction permit has been filed, as required by the FCC, within 6 months of the date of this determination. In such case, the determination expires on the date prescribed by the FCC for completion of construction, or the date the FCC denies the application.

NOTE: REQUEST FOR EXTENSION OF THE EFFECTIVE PERIOD OF THIS DETERMINATION MUST BE E-FILED AT LEAST 15 DAYS PRIOR TO THE EXPIRATION DATE. AFTER RE-EVALUATION OF CURRENT OPERATIONS IN THE AREA OF THE STRUCTURE TO DETERMINE THAT NO SIGNIFICANT AERONAUTICAL CHANGES HAVE OCCURRED, YOUR DETERMINATION MAY BE ELIGIBLE FOR ONE EXTENSION OF THE EFFECTIVE PERIOD.

Additional wind turbines or met towers proposed in the future may cause a cumulative effect on the national airspace system. This determination is based, in part, on the foregoing description which includes specific coordinates and heights . Any changes in coordinates will void this determination. Any future construction or alteration requires separate notice to the FAA.

This determination does include temporary construction equipment such as cranes, derricks, etc., which may be used during actual construction of the structure. However, this equipment shall not exceed the overall heights as indicated above. Equipment which has a height greater than the studied structure requires separate notice to the FAA.

This determination concerns the effect of this structure on the safe and efficient use of navigable airspace by aircraft and does not relieve the sponsor of compliance responsibilities relating to any law, ordinance, or regulation of any Federal, State, or local government body.

Any failure or malfunction that lasts more than thirty (30) minutes and affects a top light or flashing obstruction light, regardless of its position, should be reported immediately to (877) 487-6867 so a Notice to Airmen (NOTAM) can be issued. As soon as the normal operation is restored, notify the same number.

If we can be of further assistance, please contact our office at (404) 305-7081. On any future correspondence concerning this matter, please refer to Aeronautical Study Number 2011-WTE-6399-OE.

Signature Control No: 142406540-146136823   (DNE -WT)
Michael Blaich
Specialist

Notice of Proposed Construction or Alteration - Off Airport

Project Name: JOE W-0001745O1-11   **Sponsor:** Joe White

Details for Case : Small Wind Turbine Study

Show Project Summary

Case Status

ASN:	2011-WTE-6400-OE
Status:	Determined

Date Accepted:	05/10/2011
Date Determined:	07/14/2011
Letters:	07/14/2011 ☐ DNE
Documents:	05/10/2011 ☐ FORT DRUM SMALL W...

Structure Summary

Structure Type:	Wind Turbine
Structure Name:	Small Wind Turbine Study
NOTAM Number:	
FCC Number:	
Prior ASN:	

Construction / Alteration Information

Notice Of:	Construction
Duration:	Permanent
If Temporary :	Months: Days:
Work Schedule - Start:	07/01/2011
Work Schedule - End:	12/01/2011

*For temporary cranes-Does the permanent structure require separate notice to the FAA?
To find out, use the Notice Criteria Tool. If separate notice is required, please ensure it is filed.
If it is not filed, please state the reason in the Description of Proposal.*

State Filing:

Structure Details

Latitude:	44° 3' 9.90'' N
Longitude:	75° 43' 50.94'' W
Horizontal Datum:	NAD83
Site Elevation (SE):	675 (nearest foot)
Structure Height (AGL):	55 (nearest foot)

*If the entered AGL is a proposed change to an
existing structure's height include the current
AGL in the Description of Proposal.*

Requested Marking/Lighting:	Red lights
Other :	None
Recommended Marking/Lighting:	

Common Frequency Bands

Low Freq	High Freq	Freq Unit

Specific Frequencies

faa.g

☐ Internet | Protected Mode: On

Mail Processing Center
Federal Aviation Administration
Southwest Regional Office
Obstruction Evaluation Group
2601 Meacham Boulevard
Fort Worth, TX 76137

Aeronautical Study No.
2011-WTE-6400-OE

Issued Date: 07/14/2011

Joe White
Joe White
Bldg P-2065 Room 155 Hangar Access Drive
Fort Drum, NY 13602

** DETERMINATION OF NO HAZARD TO AIR NAVIGATION **

The Federal Aviation Administration has conducted an aeronautical study under the provisions of 49 U.S.C., Section 44718 and if applicable Title 14 of the Code of Federal Regulations, part 77, concerning:

Structure:	Wind Turbine Small Wind Turbine Study
Location:	Great Bend, NY
Latitude:	44-03-09.90N NAD 83
Longitude:	75-43-50.94W
Heights:	55 feet above ground level (AGL)
	730 feet above mean sea level (AMSL)

This aeronautical study revealed that the structure does not exceed obstruction standards and would not be a hazard to air navigation provided the following condition(s), if any, is(are) met:

Based on this evaluation, marking and lighting are not necessary for aviation safety. However, if marking/lighting are accomplished on a voluntary basis, we recommend it be installed and maintained in accordance with FAA Advisory circular 70/7460-1 K Change 2.

While the structure does not constitute a hazard to air navigation, it would be located within or near a military training area and/or route.

This determination expires on 01/14/2013 unless:

(a) extended, revised or terminated by the issuing office.
(b) the construction is subject to the licensing authority of the Federal Communications Commission (FCC) and an application for a construction permit has been filed, as required by the FCC, within 6 months of the date of this determination. In such case, the determination expires on the date prescribed by the FCC for completion of construction, or the date the FCC denies the application.

NOTE: REQUEST FOR EXTENSION OF THE EFFECTIVE PERIOD OF THIS DETERMINATION MUST BE E-FILED AT LEAST 15 DAYS PRIOR TO THE EXPIRATION DATE. AFTER RE-EVALUATION OF CURRENT OPERATIONS IN THE AREA OF THE STRUCTURE TO DETERMINE THAT NO SIGNIFICANT AERONAUTICAL CHANGES HAVE OCCURRED, YOUR DETERMINATION MAY BE ELIGIBLE FOR ONE EXTENSION OF THE EFFECTIVE PERIOD.

Additional wind turbines or met towers proposed in the future may cause a cumulative effect on the national airspace system. This determination is based, in part, on the foregoing description which includes specific coordinates and heights . Any changes in coordinates will void this determination. Any future construction or alteration requires separate notice to the FAA.

This determination does include temporary construction equipment such as cranes, derricks, etc., which may be used during actual construction of the structure. However, this equipment shall not exceed the overall heights as indicated above. Equipment which has a height greater than the studied structure requires separate notice to the FAA.

This determination concerns the effect of this structure on the safe and efficient use of navigable airspace by aircraft and does not relieve the sponsor of compliance responsibilities relating to any law, ordinance, or regulation of any Federal, State, or local government body.

Any failure or malfunction that lasts more than thirty (30) minutes and affects a top light or flashing obstruction light, regardless of its position, should be reported immediately to (877) 487-6867 so a Notice to Airmen (NOTAM) can be issued. As soon as the normal operation is restored, notify the same number.

If we can be of further assistance, please contact our office at (404) 305-7081. On any future correspondence concerning this matter, please refer to Aeronautical Study Number 2011-WTE-6400-OE.

Signature Control No: 142406612-146137053   (DNE -WT)
Michael Blaich
Specialist

Appendix 3: Letter from Fort Drum Wind Energy Working Group to Department of Defense Energy Siting Clearinghouse

IMNE-DRM-PAI

13 April 2011

INFORMATION PAPER

SUBJECT: Wind Development on the Fort Drum Installation Cantonment Area, New York. (Cold Regions Research Engineering Laboratory (CRREL) Research Wind Turbine Proposal.

1. PURPOSE: To provide information to the Department of Defense's Energy Siting Clearinghouse addressing local concerns in relation to safety of flight, impacts to training, and operational detriments. The 2011 National Defense Authorization Act (NDAA) mandated under section 358, that the Secretary of Defense would within 30 days of enactment designate a senior official at the Department of Defense to oversee a clearinghouse to review projects, and develop planning tools necessary to determine the acceptability of obstructions in the vicinity of military installations. The NDAA also directed the designated lead organization to, within 180 days assess the likely scope and duration of any adverse impacts on military operations and readiness. The Fort Drum Wind Energy Working Group (FDWEWG) assembled to discuss all the considerations of local proposed wind developments addressing safety of flight, impacts to training, and operational detriments.

2. INFORMATION:

a. In 2010, Dr Charles Ryerson from the Army's Cold Regions Research and Engineering Laboratory (CRREL), offered to conduct a test demonstration prior to fielding, two different types of small wind turbines and research the specific issues of radar interference, and environmental protection issues related to protected birds and bats. Fort Drum agreed to support the Army's research efforts, and act as the host site. A proposal was submitted to the ACSIM's research program, Installation Technology Transition Program (ITTP), for funding. The ASCIM considered research into these issues integral to accomplishing the Army's goal of incorporating renewable energy onto its installations, and subsequently funded $475,000 for the Fort Drum research effort. Fort Drum is currently researching the possibility of partnering with Clarkson University to evaluate the feasibility of small wind turbine utilization on Army Installations. The proposal's current developmental status is in the review stage, and awaiting an Environmental Impact Statement, Federal Aviation Administration (FAA) review, and United States Fish and Wildlife Service (USFWS) approval. The turbines are expected to be constructed in the summer of FY11, with testing beginning in FY12.

b. The location of the proposed research turbines would be adjacent to US Route 26, Fort Drum, Jefferson County, in the vicinity of two existing vertical obstructions with the highest obstructions height at 224 feet. The proposed turbines (tip to ground) would be approximately 34 meters or 112 feet high and fall in the shadow plane of the existing vertical obstructions, ensuring no additional vertical obstructions, thus safeguarding Fort Drum Aviators. The proposed site is approximately one mile from Fort Drum's Primary Radar.

c. As described in the attached CRREL proposal, CRREL will extensively evaluate the compatibility of small wind turbines on Army installations at Fort Drum, New York. The demonstration turbines will consist of one ~5 kW horizontal axis wind turbine, and one ~5kW vertical axis turbine. CRREL will measure turbine radar signatures, acoustic and seismic signatures, bird and bat kills, cold tolerance, accumulation of ice and snow on turbine blades, and their effects on turbine efficiency, including hazards due to ice-throw. CRREL will investigate turbine efficiency in the turbulent semi-urban environment of Fort Drum, and apply these findings throughout the Army. Demonstration turbines will be connected to the base power grid and monitored for daily power output. The proposed study will indicate how small turbines are, or are not, compatible with Army installations and their activities. The research will also address the compatibility of large numbers of turbines with installations and their activities. A product of this research will additionally provide a draft unified facilities criteria for small wind turbines in Army urban (cantonment) areas, and will be available Army-wide.

d. Due to the location and Army-wide benefits of this proposal, the Fort Drum Wind Energy Working Group (FDWEWG) does not feel there would be a significant impact to the safety of rotary-wing, or fixed wing crews training in the vicinity of WSAAF. The extensive input Fort Drum has had in the development of this proposal indicates intelligent siting restrictions, ensuring minimal impact to training and readiness. The attached CRREL proposal discusses the possible effects of the proposed turbines on the Fort Drum Wheeler-Sack Army Airfield (WSAAF) primary radar, and the NOAA NEXRAD Doppler weather radar. The CRREL proposal also will address bird and bat kills, which will be monitored by the Fort Drum Directorate of Public Works (DPW) Environmental Staff.

3. CONCLUSION:

a. The FDWEWG finds that the proposed CRREL research project would not likely have a significant impact to the safety and training ability of flight crews utilizing WSAAF. The research project will help the Army both nationally and internationally, by indicating how compatible small wind turbines can be integrated into Army Installation operations, in places where large turbines cannot be utilized due to safety of flight, and unknown radar anomaly concerns. The proposal will allow facility and acquisition managers to determine whether to adopt small wind turbines, and possibly prevent wasteful spending on wind turbine installations which are not compatible with DoD operations.

b. Fort Drum has demonstrated a consistent and prudent approach to wind turbine reviews, paying particular attention to aviation safety, encroachment, and potential impact on the Air Traffic Control (ATC) radar. Fort Drum is determined to develop wind energy in a way that ensures that we are cognizant of our off-post neighbors' concerns, while also continuing to protect the readiness and viability of Fort Drum mission.

Prepared By: MICHAEL RICHARDSON/315-772-7483
Approved By: MICHAEL H. MCKINNON/315-772-5501